Nutritional
Freshwater
Life

CONTENTS

PREFACE

Today, fishing is considered to be the largest extractive use of wildlife in the world. Freshwater fisheries provide food and a livelihood for millions of the poorest people, and also contribute to the overall economic well-being by means of export commodity trade, tourism, and recreation. It is estimated that freshwater fishes make up more than 6% of the world's annual animal protein supplies for humans. It is the major and often only source of animal protein for low-income families. Global freshwater supports at least 100,000 species out of approximately 1.8 million—almost 6% of all described species. Inland waters and freshwater biodiversity constitute a valuable natural resource in economic, cultural, aesthetic, scientific, and educational terms. It is always important to know the nutritional facts of freshwater flora and fauna occupying different habitats.

Although several books are available on limnology and aquatic biology, no books on the nutritional aspects of edible freshwater life are available presently. Keeping this in consideration, an attempt has now been made. This book deals with nutritional facts of different groups of edible freshwater life, viz., algae and plants, crustaceans (prawns, crayfish, and crabs), molluscs (bivalves and gastropods), fish, and frogs, along with their characteristics, such as classification, common names, habitats, global distribution, and biological features. When published, this book will be of great use for the undergraduate and postgraduate students of disciplines such as fisheries science, aquatic biology, biotechnology, environmental science, and life sciences, besides serving as a standard reference in the libraries of colleges and universities.

I am highly indebted to Dr. P. Velayutham, professor and head of the Department of Fish Processing Technology, Fisheries College and Research Institute, Tamilnadu Fisheries University, Thoothukudi, India, for his valuable comments. I also sincerely thank Mrs. Albin Panimalar Ramesh for her help in photography.

Ramasamy Santhanam

ABOUT THE AUTHOR

Dr. Ramasamy Santhanam is the former dean of Fisheries College and Research Institute, Tamilnadu Veterinary and Animal Sciences University, India. His fields of specialization are aquatic biology and fisheries environment. He was the principal investigator for research schemes relating to freshwater nutritional life, viz., energy and nitrogen balance in the nutrition and growth of Indian major carps (Indian Council of Agricultural Research, New Delhi, India) and nutrient chemistry and proper utility of sewage (plan scheme of Tamilnadu Veterinary and Animal Sciences University). He served as an expert for the Environment Management Capacity Building, a World Bank–aided project of the Department of Ocean Development, India. He has been a member of the American Fisheries Society, United States; World Aquaculture Society, United States; Global Fisheries Ecosystem Management Network (GFEMN), United States; and the International Union for Conservation of Nature's (IUCN) Commission on Ecosystem Management, Switzerland. To his credit, Dr. Santhanam has 12 books on fisheries science and 70 research papers.

I

INTRODUCTION

Freshwater habitats such as ponds, lakes, streams, rivers, bogs, swamps, and marshes are important components of the world. The diversity of organisms in these biotopes is important both ecologically and economically. Freshwater biomes are large communities of plants and animals centered around water with less than 1% salt concentration. Aquatic plants, such as algae, diatoms, and rooted macrophytes, use nutrients, light, and carbon dioxide to produce oxygen that, once released, is dissolved directly into the water. They also remove unwanted toxins found in the water, which could be harmful to the surrounding fish. The structure and composition of benthic communities in freshwater ecosystems is an excellent bioindicator of pollution and habitat quality.

PLANTS AND ANIMALS IN FRESHWATER BIOMES

There are many diverse flora and fauna in freshwater biomes. Algae, although not very pretty or fun to swim in, is a favorite snack for most of the animals that live in a freshwater biome. Plants are not just a snack for the freshwater animals. They provide oxygen through the process of photosynthesis. There is no shortage of animals or plants living in a freshwater biome. It is believed that more than 700 species of fish and 1200 species of amphibians, molluscs, and insects all live in these areas. There are plenty of macro animals that live in the water of the freshwater biome. They include crabs, shrimps, frogs, and turtles. Fish are very common in the freshwater biome. However, where there is freshwater, one can often find a huge number of different types of mammals.

MALNUTRITION, A GLOBAL PROBLEM

At present, more than 30% of humanity is suffering from malnutrition and food-related diseases, either in the form of malnutrition and undernourishment or in the form of excessive nutrient intake and obesity. The global reach of malnutrition and food insecurity is such that hunger is still the world's number one health risk, killing more people every year than acquired immunodeficiency syndrome (AIDS), malaria, and tuberculosis combined (Tacon et al., 2010; World Food Programme (WFP), 2012). The global magnitude and consequences of hunger and malnutrition are profound and long-lasting, with 925 million chronically undernourished people within the developing world and over 6.6 million child deaths every year

1

related to malnutrition. Further, more than 2 billion people in the world suffer from specific dietary micronutrient deficiencies, including iron, iodine, vitamin A, and zinc (WFP, 2012). In terms of per capita food supply, total aquatic animal food supply (live weight equivalent) has grown from 11.1 kg in 1970 to 18.6 kg in 2010, and per capita aquatic meat supply has grown from 8.7 kg in 1970 to 13.2 kg in 2010 (FAO/FAOSTAT, 2012).

IMPORTANCE OF FRESHWATER FISHERIES

Today, fishing remains the largest use of wildlife in the world. Ninety-four percent of all freshwater fisheries occur in developing countries, and freshwater fishes make up more than 6% of the world's annual animal protein supplies for humans (FAO, 2007). They provide food and a livelihood for millions of the world's poorest people, and also contribute to the overall economic well-being by means of export commodity trade, tourism, and recreation. Some people derive great aesthetic pleasure from recreational activities in freshwaters, such as fishing, swimming, kayaking, and canoeing. Recreation and tourism help the entire industry of skilled craftspeople.

IMPORTANCE OF FRESHWATER FISH

Freshwater fish make up an important portion of aquatic organisms that are used as human foodstuffs. In many regions of the world, freshwater fish are a significant source of animal protein. They also contain considerable quantities of valuable lipids as well as minerals and vitamins.

PROTEIN CONTENT

In the dry matter of muscle or flesh, protein is the main component. The protein content generally amounts to 15 to 20% of wet weight in the muscle. In lipid-rich fish species, the protein level is lower than in species with poor lipid content in their flesh. The amino acid composition of the protein in different freshwater fish species is more or less similar.

LIPID CONTENT

Depending on species, size, and nutrition, the lipid level of the flesh in freshwater fish can vary considerably. There are freshwater fish species like pike (*Esox lucius*), perch (*Perca fluviatilis*), and pike-perch (*Sander lucioperca*) with exceptionally lean meat, while other species are characterized by medium or even high lipid levels in their muscle, e.g., eel (*Anguilla anguilla*). Lipid content varies between less than 5% and more than 50% in the dry matter of muscle. With growing fish size, the lipid level in the muscle generally increases. The lipid content of large rainbow trout (*Oncorhynchus mykiss*) weighing more than 1000 g frequently is about 10% of wet weight.

2

LIPID QUALITY

Lipids of freshwater fish are characterized by high levels of n-3 polyunsaturated fatty acids, e.g., eicosapentaenoic acid (EPA) and docosahexaenoic acid (DHA). But they also contain considerable amounts of n-6 fatty acids, especially linoleic acid and arachidonic acid. While the essential fatty acid ratio (n-3/n-6) of marine fish generally varies between 5 and 10, the essential fatty acid ratio of freshwater fish is 1 to 4. This seems appropriate for human nutrition. Because of their favorable fatty acid composition, the lipids of freshwater fish are of outstanding nutritional significance. It is well known that not only the n-6 fatty acids are essential for human nutrition, but the n-3 fatty acids are just as necessary and must also be supplied in the diet. Both series of essential fatty acids cannot be synthesized by other animals or humans. Clinical tests with hypertensive patients in many countries proved the effectiveness of eating freshwater fish in lowering blood pressure and plasma lipids. Hypertensive patients were put on a 2-week diet of 100 g silver carp meat (*Hypophthalmichthys molitrix*) per day. This resulted in a significant drop in systolic blood pressure by 15 mmHg and diastolic pressure by 9 mmHg. In blood plasma, the level of triacylglycerols was reduced by 0.6 mmol/L, the HDL cholesterol increased by 0.26 mmol/L, and the phospholipid concentration remained constant.

MINERAL AND VITAMIN CONTENT

Freshwater fish are also a good source for several minerals and vitamins in human nutrition. Their meat is rich in potassium, phosphorus, sodium, magnesium, and calcium. The flesh of freshwater fish also contains some trace elements, e.g., iron and anganese. Freshwater fish can also meet the demand for several vitamins, such as vitamin A and some B vitamins.

FRESHWATER FISH IN HUMAN NUTRITION

Freshwater fish are a nutritious choice for lunch or dinner, as they are low in fat and high in protein. It has been reported that the average freshwater fish consumption is 56.6 kg/person/year, and one serving of most freshwater fish provides more than 30% of the dietary reference intake of protein for adults. Freshwater fishes form the major and often only source of animal protein for low-income families. Although not as high in healthy omega-3 fatty acids as some saltwater fish, one serving of most freshwater fish provides more than 30% of the dietary reference intake of protein for adults.

Freshwater fish are therefore considered to be wholesome foodstuffs of high nutritive value. As freshwater fish flesh has good digestibility, it is also well suited as a healthy diet for children and seniors (Steffens, 2006).

FRESHWATER CRUSTACEANS

Among freshwater crustaceans, crayfish assume greater importance, as they have a super healthy combination of nutrients. From an almost pure form of protein to a healthy amount of omega-3 fatty acids, they are the most beneficial fats to eat for human nutrition. Crayfish protein has large amounts of the amino acid tyrosine that mentally energizes the brain. In addition, there is a healthy supply of vitamins D and A, as well as calcium, potassium, copper, and zinc in crayfish. Iodine is also often mentioned as an important ingredient. Further, crayfish are a very low carbohydrate food, and you can safely eat crayfish without putting on unwanted weight. Next to crayfish, freshwater crabs are important, as their meat is nutritionally good, being high in vitamins, high-quality proteins, and amino acids. It is also rich in minerals such as calcium, copper, zinc, phosphorus, and iron, while having lower levels of fat and carbohydrates.

FRESHWATER MOLLUSCS

Freshwater bivalves, the class of molluscs that includes clams and mussels, are extremely rich in a unique combination of nutrients that promote men's health. These organisms are a superior source of low-calorie protein loaded with minerals such as potassium, phosphorus, manganese, iron, selenium, and zinc and vitamins such as B3, B12, C, and riboflavin. Three ounces of raw clams will cost you only 63 calories, but you get 11 g of protein, 66% of the daily recommended amount of iron, and 700% of the daily recommended amount of vitamin B12. Chinese medicine recommends clams for treating hemorrhoids. The importance of freshwater gastropods (snails) in mitigating the protein deficiency in poor countries, such as Bangladesh, cannot be overlooked.

FROGS

A number of species of frogs, including the bullfrog (*Rana catesbeiana*) and green frog (*Rana clamitans*), are nowadays harvested from the wild, and their legs are sold as a luxury food in expensive restaurants of several countries.

A summary of compositional data of nutritional facts of major freshwater faunal groups is given below.

Nutritional Status of Major Freshwater Groups (per 100 g (Fresh Weight Basis))

	Protein (g)		Fat (g)		Ca (mg)		Fe (mg)		Vitamin A (µg)	
	Min	Max	Min	Max	Min	Max	Min	Max	Min	Max
Fish	9.7	22.7	0.8	8	17	1751	0.6	9.2	5	1800
Crustacean	10.7	21.2	0.9	3.3	75	5000	0.6	7.5	0	133
Mollusc	7	20.2	0.3	1.4	16	2500	7	26.6	0	243
Frogs	15.1	20.5	0.2	2	19	1293	0.7	3.8	Low	

Source: Data from James (2006); Nurhasan (2008).

MICROALGAE

Algae have been used as a food source and for treatment of various ailments for over 2000 years. They form numerous compounds that are presently used in the development of new nutraceuticals and have the potential to become more intensively exploited. Different types of algae, specifically microalgae, are more prevalent in food supplements and nutraceuticals due to their capability of producing necessary vitamins, including A (retinol), B1 (thiamine), B2 (riboflavin), B3 (niacin), B6 (pyridoxine), B9 (folic acid), B12 (cobalamin), C (L-ascorbic acid), D, E (tocopherol), and H (biotin).

Also, these organisms concentrate essential elements, including potassium, zinc, iodine, selenium, iron, manganese, copper, phosphorus, sodium, nitrogen, magnesium, cobalt, molybdenum, sulfur, and calcium. Algae are also high producers of essential amino acids and omega 6 (arachidonic acid) and omega 3 (docosahexaenoic acid, eicosapentaenoic acid) fatty acids. Due to their abundant production of beneficial compounds and nutritive contents, the market for increased algae production for nutraceuticals is lucrative and imminent.

AQUATIC PLANTS

Some aquatic plants have a long tradition as human food. *Oryza sativa* is an emergent aquatic plant and the only vascular hydrophyte that is a major agronomic species. It is one of the world's main crops and forms the staple diet of over half the world's population. A small but important number of other aquatic crops exist, such as water chestnut (*Trapa* spp.), lotus (*Nelumbo nucifera*), and watercress (*Nasturtium officinale*). *Ipomoea aquatica* is one of the few aquatics grown as a green vegetable.

Aquatic macrophytes present organs that, due to their accumulated food reserves, are of potential nutritional value to man: among them, seeds, fruits, and swollen vegetative perennating organs are the most important. A variety of fruits and seeds are rich in oil, starch, or protein and can be eaten raw or dried and ground to flour, which can be baked with water or milk to give a kind of bread or cake. Numerous rhizomes and tubers are similarly rich in carbohydrates, especially starch, sugar,

and mucilage, and are wholly edible when raw or cooked. The foliage of many macrophytes provides acceptable salad ingredients or cooked vegetable dishes. In the Amazon, Indians used to utilize water hyacinth ashes as salt, and in the Northeast region of Brazil, the leaves of several species of macrophytes are often used in salads due to their high nutritional value and good palatability. The rhizomes of many other macrophytes are also used in the production of cookies, cakes, and other products. In Brazil, one of the most used macrophytes with nutritional value to man is the watercress (*Nasturtium* sp.), and it is often used in fresh or cooked dishes (Thomaz et al., 2009).

In general, aquatic plants are somewhat low in lysine and methionine when compared to meat proteins. Phenylalanine is present in very small quantities in *Alternanthera philoxeroides* and *Sagittaria latifolia*. Otherwise, amino acid levels are similar to those for meat proteins and crop plant leaf protein isolates. Hence, leaf protein from aquatic plants is of sufficiently high amino acid quality to be useful as a dietary supplement. As the nutritive value of aquatic plants often contains "as much or more crude protein, crude fat, and mineral matter, than many conventional forage crops," such plants could be used as food and temporarily help alleviate food shortages until lasting agricultural and social solutions are found (Boyd, 1968a, 1968b).

In order to alleviate the problems relating to malnutrition among the people of developing and underdeveloped countries, there is an urgent need to identify new species rich in nutrients. In this regard, assessment of the nutritional quality of freshwater flora and fauna of edible value would be of great use to add new and cheap sources of animal proteins. Further, by applying the knowledge of nutritional status, one can select the needed species to harvest. This may help conserve the freshwater ecosystems and their biodiversity, which is presently an urgent need (Virginia Tech, 2009). In the present report, an attempt has been made to present detailed information on the nutritional facts of different groups of freshwater life. All table information relates to mean values of fresh weight (FW) (wet weight (WW)) or dry matter (DM) (dry weight (DW)) adult specimens.

2
ALGAE AND PLANTS

ALGAE

PHYLUM: CHLOROPHYTA (GREEN ALGAE)

Chlorella vulgaris (Beyerinck) Beijerinck 1890

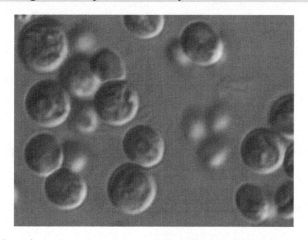

Phylum: Chlorophyta

Class: Trebouxiophyceae

Order: Chlorellales

Family: Chlorellaceae

Common name: Chlorella.

Distribution: Atlantic islands, North America, South America, Southwest Asia, Southeast Asia, Australia and New Zealand, Europe (AlgaeBase).

Habitat: Freshwater and terrestrial habitats.

Description: This species is small, unicellular, and nonmotile, about 2–15 µm in diameter. Cells are spherical or ellipsoidal with a single, parietal, cup-shaped (sometimes plate-like) chloroplast with or without a pyrenoid. Cell wall is generally thin and smooth. Cells contain green photosynthetic pigments, viz., chlorophyll *a* and *b* in the chloroplast. It is an attractive potential food source because it is high in protein and other essential nutrients. It is also a popular dietary supplement (van Vuuren et al., 2006).

Nutritional Facts

Proximate Composition (g/100 g DM)

Moisture	Ash	Protein	Lipid	CHO[a]	Fiber
9.95	6.87	48.19	5.60	29.85	17.06

[a] Carbohydrate.

Minerals (mg/100 g)

Ca	Mg	Zn	Na	K	Cu	Fe
1425	851	293	101	1197	48	82

Vitamins

Retinol (µg/g)	C (mg/g)	E (mg/g)
132.15	0.39	27.87

Source: Data from Yusof et al. (2011).

Amino Acids (g/16 gN)

Isoleucine	3.8
Leucine	8.8
Valine	5.5
Lysine	8.4
Phenylalanine	5.0
Tyrosine	3.4
Methionine	2.2
Cysteine	1.4

Tryptophan	2.1
Threonine	4.8
Alanine	7.9
Arginine	6.4
Aspartic acid	9.0
Glutamic acid	11.6
Glycine	5.8
Histidine	2.0
Proline	4.8
Serine	4.1

Fatty Acids (% of Total Lipids)

12:0	0
14:0	0.9
14:1	2.0
15:0	1.6
16:0	20.4
16:1	5.8
16:2	1.7
17:0	2.5
18:0	15.3
18:1	6.6
18:2	1.5
20:2	1.5
20:3	20.8

Source: Data from Becker (1994).

Chlorella pyrenoidosa H. Chick 1903

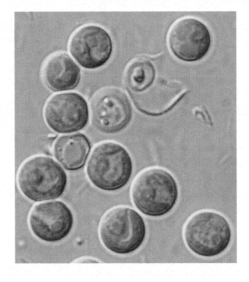

Order: Chlorellales

Family: Chlorellaceae

Common name: Chlorella.

Distribution: Worldwide; Europe, Asia, Australia and New Zealand (AlgaeBase).

Habitat: Freshwater bodies.

Description: This species is small, unicellular, and nonmotile, about 2–15 µm in diameter. Cells are spherical or ellipsoidal with a single, parietal, cup-shaped (sometimes plate-like) chloroplast with or without a pyrenoid. Cell wall is generally thin and smooth. Cells contain green photosynthetic pigments, viz., chlorophyll *a* and *b* in the chloroplast. Over 24 vitamins and minerals (plus vital trace elements) and 19 amino acids (including all 8 essential ones) are available in this species. Further, highly digestible, complete protein and essential fatty acids—omega-3, omega-6, and gamma-linolenic acid (GLA)—are also present. Hence, this species is considered a popular food supplement (van Vuuren et al., 2006).

Nutritional Facts

Proximate Composition (% DM)

Protein	CHO[a]	Lipids
57	26	2

[a] Carbohydrate.

Source: Data from Becker (1994).

Amino Acids (g/16gN)

Isoleucine	3.4
Leucine	4.0
Valine	5.1
Lysine	7.9
Phenylalanine	4.5
Tyrosine	2.7
Methionine	1.8
Cysteine	0
Tryptophan	1.4
Threonine	3.2
Alanine	5.9
Arginine	5.6
Aspartic acid	5.9
Glutamic acid	9.3
Glycine	4.8
Histidine	1.4
Proline	4.0
Serine	2.2

Source: Data from Becker (1994).

Chlorella ellipsoidea Gernceck 1907

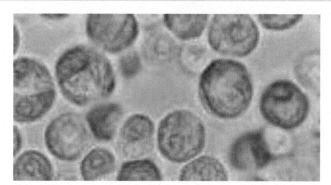

Order: Chlorellales

Family: Chlorellaceae

Common name: Chlorella.

Distribution: Europe, Southwest Asia, Caribbean islands, Australia and New Zealand (AlgaeBase).

Habitat: Freshwater and terrestrial areas.

Description: Cells of this species are 1.5–13 µm wide and 2–15 µm long. Chloroplast is more lobed with age. Margins are irregularly undulate and occasionally incised. Pyrenoid is associated with numerous starch grains. Autospores are ellipsoidal. This species serves as a food supplement (van Vuuren et al., 2006).

Nutritional Facts

Proximate Composition (% DM)

Protein	Fiber	Ash	Moisture
22.39–39.91	3.66–5.19	10.10–11.11	8.15–9.45

Source: Data from Mondal et al. (2005).

Amino Acids (g/16 gN)	
Isoleucine	4.5
Leucine	9.3
Valine	7.9
Lysine	5.9
Phenylalanine	4.2
Tyrosine	1.7
Methionine	0.6
Cysteine	0.7
Threonine	4.9
Alanine	12.2
Arginine	5.8
Aspartic acid	8.8
Glutamic acid	10.5
Glycine	10.4
Histidine	1.7
Proline	5.0
Serine	5.2

Source: Data from Becker (1994).

Prasiola japonica Yatabe 1891

Order: Prasiolales

Family: Prasiolaceae

Common name: River nori.

Distribution: Asia: China, Japan (AlgaeBase).

Habitat: Freshwater habitat, river streams.

Description: Thalli of this species are up to 0.5 cm long and consist of bright to dark green monostromatic blades with an irregularly rounded, ovate shape. Blades, devoid of stripes, are attached to the substratum by means of a rim of the frond or free and are 17–20 μm thick. Thalli have a smooth to irregular margin. Each cell has a stellate plastid with a pyrenoid in the center. Nuclear structures are primitive with little differentiation, and there are no motile reproductive cells. It is a food species (Iwamoto, 1984).

Nutritional Facts

Proximate Composition (g/100 g DM)

Protein	Fat	CHO[a]	Ash
42.03	1.76	45.9	4.7

[a] Carbohydrate.

Minerals (g/100 g DM)

Ca	P	Fe[a]
1.03	0.7	99

[a] mg%.

Source: Data from Johnston (1970).

Cladophora sp.

Phylum: Chlorophyta

Class: Ulvophyceae

Order: Cladophorales

Family: Cladophoraceae

Common name: Mekong weed.

Distribution: Cosmopolitan in temperate and tropical regions (AlgaeBase).

Habitat: Attached to rocks or timbers submerged in shallow lakes and streams; freshwater, brackish, and marine conditions.

Description: This unidentified species is reticulated and filamentous. It is coarse in appearance, with regular branching filaments that have cross-walls separating multinucleate segments. There are two multicellular stages in its life cycle—a haploid gametophyte and a diploid sporophyte—which look highly similar. It grows in the form of a tuft or ball with filaments that may range up to 13 cm (5 in.) in length. If attached, branching rhizoids arise from basal cells and other cells in the basal region, or a simple discoid holdfast is produced. Chloroplasts are parietal, either densely packed discoid or united in a reticulum. Cells are multinucleate. This species is edible (AlgaeBase, 2000).

Nutritional Facts

Proximate Composition (% DM)

Protein	Lipid	Ash	Moisture	Fiber	CHO[a]
10.71	2.04	15.29	10.98	23.05	60.98

[a] Carbohydrate.

Vitamins (mg/100 g)

A	B1	B2	C	E
0.33	0.05	0.05	1.89	5.97

Minerals (mg/100 g)

Ca	Fe	K	Mg	P	Zn
6.631	29.91	2.658	241.60	69.17	1.91

Carotenoid (μg/g)

Beta-carotene	Lutein	Zeaxanthin
20.01	172.80	24.62

Source: Data from Khuantrairong and Traichaiyaporn (2011).

Cladophora glomerata (Linnaeus) Kützing 1843

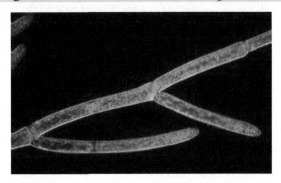

Order: Cladophorales

Family: Cladophoraceae

Common name: Mekong weed.

Distribution: Throughout North America, Europe, the Atlantic islands, the Caribbean islands, Asia, Africa, Australia and New Zealand, and the Pacific islands (AlgaeBase).

Habitat: Marine/freshwater species.

Description: Filaments of this species are branched and are attached by rhizoids. They are dark green and grow up to 200 cm. Branching is pseudodichotomous with insertion of side branches generally oblique to horizontal. Cells are mostly cylindrical, 3–30 times longer than wide. Chloroplast is parietal net-like with several pyrenoids that are composed of two halves (bilenticular). Apical cells are cylindrical, often slightly tapering, 21–90 µm wide, and 1.5–13 times longer than wide. This species is a great example of a dietary supplement (Ramin, 2013; LRMW, 2001).

Nutritional Facts

Proximate Composition (% DM)

Protein	CHO[a]	Lipid
20.38	14.83	1.1

[a] Carbohydrate.

Source: Data from Manivannan et al. (2009).

Scenedesmus obliquus (Turpin) Kützing 1833 (= *Scenedesmus dimorphus*)

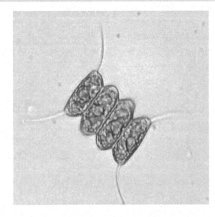

Phylum: Chlorophyta
Class: Chlorophyceae
Order: Sphaeropleales
Family: Scenedesmaceae
Common name: Green alga.
Distribution: Found across the world (AlgaeBase).
Habitat: Ponds, lakes, and rivers.

Description: Cells of this species commonly occur in colonies (coenobia) as multiples of two, with four or eight cells being most common. Coenobia are formed inside the mother cell wall and are released to either form a new colony or disperse as unicells, which are elliptical, measuring 10 μm long. Different forms of coenobia are found, including linear, costulatoid, irregular, alternating, and dactylococcoid patterns. Uninucleate cells have a single laminate chloroplast, which constitutes most of the cell volume. Cell wall contains mainly cellulose, pectin, and the polycarotenoid sporopollenine, which gives it extreme resistance. This species could serve as a useful dietary supplement. It has been reported that incorporation of this species at 5 g/daily in the diet has increased weight in children (Gouveia et al., 2008; Wünchiers, 2002).

Nutritional Facts

Proximate Composition (% DM)

Protein	CHO[a]	Lipids
50–56	10–17	12–14

[a] Carbohydrate.

Amino Acids (g/100 g protein)

Isoleucine	3.6
Leucine	7.3
Valine	6.0
Lysine	5.6
Phenylalanine	4.8
Tyrosine	3.2
Methionine	1.5
Cysteine	0.6
Tryptophan	0.3
Threonine	5.1
Alanine	9.0
Arginine	7.1
Aspartic acid	8.4
Glutamic acid	10.7
Glycine	7.1
Histidine	2.1
Proline	3.9
Serine	3.8

Source: Data from Becker (2006).

Fatty Acids (% of lipids)

12:0	0.3
14:0	0.6
14:1	0.1
16:0	16.0
16:1	8.0
16:2	1.0
16:4	26.0
18:0	0.3
18:1	8.0
18:2	6.0

Source: Data from Becker (1994).

Chlamydomonas reinhardtii (P.A. Dang)

Order: Chlamydomonadales

Family: Chlamydomonadaceae

Common name: Green alga.

Distribution: Worldwide.

Habitat: Soil and freshwater.

Description: It is a single-cell green alga about 10 μm in diameter that swims with two flagella. It has a large cup-shaped chloroplast, a large pyrenoid, and an "eyespot" that senses light. Each cell typically has two anterior contractile vacuoles. Nucleus is single and typically central. Basal bodies, connected by striated (proximal and distal) fiber systems, exhibit clockwise absolute configuration. The extract of this species could be used as a substitute for yeast extract in food-related industries (Kightlinger et al., 2014).

Nutritional Facts

Proximate Composition (% DM)

Protein	CHO[a]	Lipids
48	17	21

[a] Carbohydrate.

Source: Data from Becker (2006).

Haematococcus pluvialis Flotow 1844

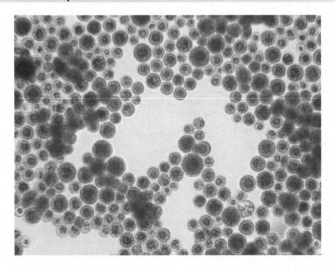

Order: Chlamydomonadales (Volvocales)

Family: Haematococcaceae

Common name: Green alga.

Distribution: Temperate regions around the world; Asia: Taiwan.

Habitat: Inland and coastal rock pools, rain pools, and bog pools.

Description: During the life cycle of this species, four types of cells are distinguished: microzooids, large flagellated macrozooids, nonmotile palmella forms, and hematocysts. Hematocysts that are formed during the resting stage are large, oval-shaped, red cells with a heavy, resistant, thickened cell wall, but no flagella. The pigment astaxanthin of this stage is largely responsible for the development of red coloration during its blooms. The hematocyst stage gets transformed into a flagellated macrozooid stage that is surrounded by a gelatinous

14

sheet and is green colored with two flagella. The latter stage develops into a palmella stage (macrozooid) that is green, spherical, and without flagella. The astaxanthin content (5.7 mg/g DW) of the cells makes them very useful as a food additive (Burchardt et al., 2006; Dragoş et al., 2010).

Nutritional Facts

Proximate Composition (DM)

	Minimum	Maximum	Mean
Protein (%)	17.30	27.16	23.62
Carbohydrates (%)	36.9	40.0	38.0
Fat (%)	7.14	21.22	13.80
Iron (%)	0.14	1.0	0.73
Moisture (%)	3.0	9.0	6.0
Ash (%)	11.07	24.47	17.71

Minerals

	Minimum	Maximum	Mean
Magnesium (%)	0.85	1.4	1.14
Calcium (%)	0.93	3.3	1.58

Vitamins

	Minimum	Maximum	Mean
Biotin (mg/lb)	0.108	0.665	0.337
L-carnitine (µg/g)	7.0	12.0	7.5
Folic acid (mg/100 g)	0.936	1.48	1.30
Niacin (mg/lb)	20.2	35.2	29.8
Pantothenic acid (mg/lb)	2.80	10.57	6.14
Vitamin B1 (mg/lb)	<0.050	4.81	2.17
Vitamin B2 (mg/lb)	5.17	9.36	7.67
Vitamin B6 (mg/lb)	0.659	4.5	1.63
Vitamin B12 (mg/lb)	0.381	0.912	0.549
Vitamin C (mg/lb)	6.42	82.7	38.86
Vitamin E (IU/lb)	58.4	333	186.1

Amino Acids (% DM)

	Minimum	Maximum	Mean
Tryptophan	0.05	0.56	0.31
Aspartic acid	1.37	2.31	1.89
Threonine	0.78	1.24	1.04
Serine	0.73	1.06	0.94
Glutamic acid	1.70	2.39	2.19
Proline	0.69	1.00	0.89
Glycine	0.84	1.32	1.17
Alanine	1.30	1.92	1.73
Cysteine	0.16	0.21	0.19
Valine	0.83	1.94	1.36
Methionine	0.32	0.43	0.40
Isoleucine	0.55	0.97	0.79
Leucine	1.21	1.84	1.67
Tyrosine	0.40	0.63	0.52
Phenylalanine	0.61	1.05	0.90
Histidine	0.48	0.76	0.61
Lysine	0.75	1.32	1.13
Arginine	0.81	1.34	1.07

Source: Data from Lorenz (1999).

Lipids (% DM)

Total Lipids	Neutral Lipids	Glycolipids	Phospholipids
15.61	9.20	3.7	1.87

Fatty Acid Profile (% of total fatty acids)

SFA	MUFA	PUFA
27.81	20.07	45.80

SFA, saturated fatty acid; MUFA, monounsaturated fatty acid; PUFA, polyunsaturated fatty acid.

Source: Data from Damiani et al. (2010).

Pigment Composition (mg/g DM)

Pigment	Green Stage	Red Stage
Chlorophyll *a*	10.053	4.027
Chlorophyll *b*	7.829	1.665
Astaxanthin	0.572	5.753
Carotene (α + β)	0.295	0.542
Zeaxanthin	0.352	0.317
Lutein	1.554	1.471
Lactucaxanthin	ND	0.258
Violaxanthin	0.274	0.241
Neoxanthin	0.338	0.512
Carotenoids/ chlorophylls	0.189	1.598

ND, no data.

Source: Data from Dragoş et al. (2010).

Phylum: Cyanophyta (Blue-Green Algae)

Nostochopsis lobatus H.C. Wood ex Bornet & Flahault 1886

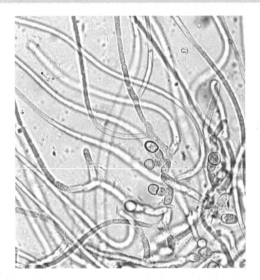

Phylum: Cyanophyta

Class: Cyanophyceae

Order: Nostocales

Family: Nostochopsidaceae

Common name: Blue-green alga.

Distribution: Europe, South America, Asia, Australia and New Zealand, Pacific islands (AlgaeBase).

Habitat: Shallow rivers or streams.

Description: This species consists of a dark green colony with a mucilaginous sheath outgrowth and is attached on rocks. Scattered branches with the lateral branch stretching parallel and upward to the main filament in a polysaccharide sheath are its unique descriptive characteristics. *N. lobatus* is of great interest as a potential source of food, as a food supplement, and as a pharmaceutical product (Thiamdao et al., 2012).

Nutritional Facts

Proximate Composition (% DM)

Lipid	Protein	CHO[a]	Fiber
0.64	19.10	31.94	2.05

[a] Carbohydrate.

16

Vitamins (mg/100 g DM)

C	B1	B2	Niacin
1.07	0.12	0.07	2.48

Minerals (mg/100 g DM)

Ca	Na	K	Cl	Mg	Mn	Fe	Zn	Se[a]
6405	136.9	0.5	0.3	265.4	4.5	114.9	0.65	37

[a] μg/100 g DW.

Pigments (mg/g Cell DM)

Chlorophyll *a*	8.26
Carotenoid	0.339
Phycocyanin	61.58
Allophycocyanin	65.38

Source: Data from Thiamdao et al. (2012).

Aphanizomenon flos-aquae (Ralfs ex Bornet & Flahault Nevada)

Order: Nostocales

Family: Nostocaceae

Common name: Blue-green algae.

Distribution: Europe, North America, Southwest Asia, Asia, Australia and New Zealand.

Habitat: Lakes, rivers, brackish waters.

Description: This species grows to form single cells (akinetes) or single, straight, and unbranched filaments that may aggregate into large spindle- or flake-like colonies (up to 3 × 30 mm), resulting in small "glass blades." Filaments without sheaths (trichomes) are tapered at both ends, with end cells usually longer than the central cells. Cells are 5–6 μm wide and 5–15 μm long with gas vacuoles. Wall of the filament is smooth and colorless. This species is used as a valuable dietary supplement (Baker, 2012; Laamanen et al., 2002).

Nutritional Facts (per g)

Proximate Composition (% DM)

Protein	60–70
CHO[a]	20–30
Energy	260[b]
Minerals	3–9
Lipids	2–8
Pigments	1–4
Moisture	3–7

[a] Carbohydrate.

[b] kcal/100 g.

17

Essential Fatty Acids (per g)

Alpha-Linolenic acid (omega-3)	29.50 mg
Gamma-Linolenic acid (omega-6)	6.00 mg

Vitamins (per g)

Provitamin A beta-carotene	2000 IU
Vitamin E	1.70 IU
Thiamin (B1)	4.70 µg
Ascorbic acid (vitamin C)	6.70 mg
Riboflavin (B2)	57.30 µg
Biotin	0.30 µg
Niacin (B3)	0.16 mg
Folic acid	1.00 µg
Pantothenic acid (B5)	6.80 µg
Choline	2.30 µg
Pyridoxine (B6)	11.10 µg
Cobalamin (B12)	8.00 µg
Inositol	160 µg
Vitamin K	45.52 µg

Minerals (per g)

Boron	0.15 mg
Iodine	0.53 µg
Selenium	0.67 µg
Calcium	14.00 mg
Iron	350.70 µg
Silicon	186.50 µg
Chloride	0.47 mg
Magnesium	2.20 mg
Sodium	2.70 mg
Chromium	0.53 µg
Manganese	32.00 µg
Tin	0.47 µg
Cobalt	2.00 µg
Molybdenum	3.30 µg

Titanium	46.60 µg
Copper	4.30 µg
Nickel	5.30 µg
Vanadium	2.70 µg
Fluoride	38.00 µg
Potassium	12.00 µg
Zinc	18.70 µg
Germanium	0.27 µg
Phosphorus	5.20 µg

Amino Acids (per g)

Essential Amino Acids	
Arginine	38 mg
Methionine	7 mg
Histidine	9 mg
Phenylalanine	25 mg
Isoleucine	29 mg
Threonine	33 mg
Leucine	52 mg
Tryptophan	7 mg
Lysine	35 mg
Valine	32 mg
Nonessential Amino Acids	
Alanine	47 mg
Glutamine	78 mg
Asparagine	47 mg
Glycine	29 mg
Aspartic acid	7 mg
Proline	29 mg
Cystine	2 mg
Serine	29 mg
Glutamic acid	4 mg
Tyrosine	17 mg

Source: Data from Bluegreen Foods (2001–2014).

Anabaena cylindrica **Lemmermann 1896**

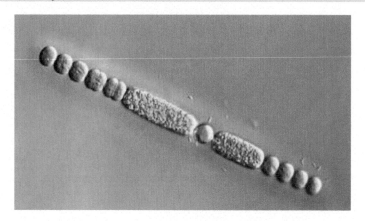

Order: Nostocales

Family: Nostocaceae

Common name: Not available.

Distribution: Europe, Asia, Australia and New Zealand, North America (AlgaeBase).

Habitat: Ponds, shallow water, and on moist soil.

Description: Trichomes are straight and are often aggregated to form a blue-green colony. Mucilage is thin and usually colorless, but sometimes with a pale yellow-brown tinge. Mucilage is more evident in aged material, especially around heterocysts and akinetes. Cells are subspherical or ellipsoidal, measuring 3–4 µm wide and 5–7 µm in length. Akinetes are cylindrical,

5–8 µm wide, and 21–30 µm long. The polysaccharides produced by this species may be of great use in medical applications (Bishop et al., 1954).

Nutritional Facts

Proximate Composition (% DM)

	VW	HW	SW	SM
CHO[a]	18	62	41	66
Amino compounds	65	4	24	5
Lipid	3	15	11	0
Ash	2.5	2	2	7
Moisture	3	2.5	2	8

VW, vegetative wall; HW, heterocyst wall; SW, spore wall; SM, sheath material.

[a] Carbohydrate.

Source: Data from Dunn and Wolk (1970).

19

Calothrix fusca Bornet & Flahault 1886

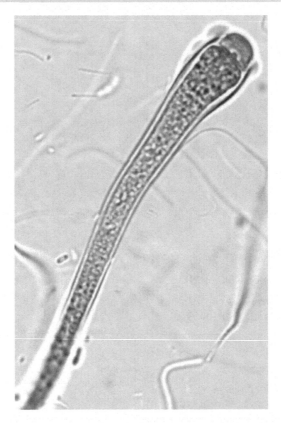

Order: Nostocales

Family: Rivulariaceae

Common name: Blue-green alga.

Distribution: Arctic, Ireland, Europe, North America, South America, Southwest Asia, Australia and New Zealand, Pacific islands (AlgaeBase).

Habitat: Freshwater/terrestrial species.

Description: Filaments are single or in small groups. They are unbranched or with occasional false branching. Trichome is straight or bent, ending in a distinct hair. Cells near the base measure 7–8 μm wide. Sheath is colorless.

This species is a dietary supplement (John, 2002).

Nutritional Facts

Proximate Composition (% DM)

CHO[a]	Protein	Amino Acid	Lipid
18.5	1.6	2.5	10.5

[a] Carbohydrate.

Minerals (μg/ml)

Ca	Mn	Fe	Zn	Ni	Mg
38.9	204.8	4779	94.0	13.2	13,350

Source: Data from Rajeshwari and Rajashekhar (2011).

Scytonema bohneri Schmidle 1901

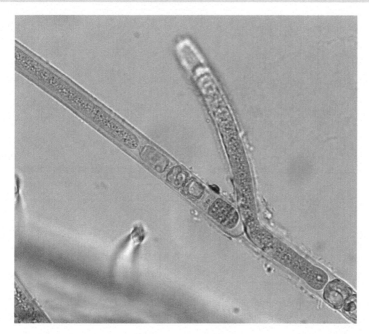

Order: Nostocales

Family: Scytonemataceae

Common name: Not available.

Distribution: Throughout the tropical regions (Komárek et al., 2013).

Habitat: Freshwater habitats; thermal waters.

Description: Filaments of this species are blackish green, partly creeping, partly ascending. Double false branching between two heterocysts and seldom single false branching may be seen. Trichomes are blue-green to greenish; cells 6–8.5 µm wide, usually isodiametric, shorter at the apices, slightly longer than wide in the rest. Heterocyst is rounded quadrate, or ellipsoidal, single or many, 7–8.9 µm wide, up to 16 µm long. Sheaths are firm, up to 2 µm thick, typically colorless, sometimes yellow, rarely lamellate. It is an edible microbe with a high food value (SBSAC).

Nutritional Facts

Proximate Composition (% DM)

CHO[a]	Protein	Amino Acid	Lipid
28.5	0.7	3.0	11.0

[a] Carbohydrate.

Minerals (µg/ml)

Ca	Mn	Fe	Zn	Ni	Mg
43.0	211.5	3367	102.6	11.05	13,190

Source: Data from Rajeshwari and Rajashekhar (2011).

Arthrospira platensis Gomont 1892 (= *Spirulina platensis*)

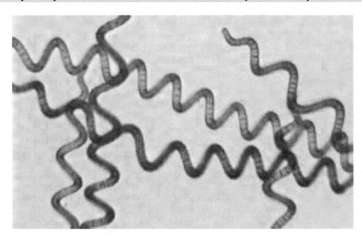

Order: Chroococcales

Family: Spirulinaceae

Common name: Spirulina.

Distribution: Tropical and subtropical bodies of water; Africa, Asia, and South America (AlgaeBase).

Habitat: Lakes.

Description: It is a free-floating species that tends to aggregate and form mats along the periphery of the aqueous environment. Trichome is approximately 5 μm in diameter. Trichomes vary greatly in length and are typically spiral shaped, though they can take on a left-hand helical structure in liquid media. When the trichome is mature, it breaks up into short cellular chains of two to four cells, or hormogonia, which glide away and begin new trichomes. Individual cells within the filament are wider than they are long and are separated by transverse cross-walls. *A. platensis* does not possess any flagella. This serves as a useful food supplement (Madkour et al., 2012; djwesten@mst.edu).

Nutritional Facts

Proximate Composition (% DM)

Protein	CHO[a]	Lipid
52.62	15.00	6.50

[a] Carbohydrate.

Source: Data from Madkour et al. (2012).

Fatty Acids (% DM)

C16:0	C16:1	C18:0	C18:1	C18:2	C18:3
45.92	2.74	0.89	7.77	11.95	20.63

Note: C16:0, palmitic acid; C16:1, palmitoleic acid; C18:0, stearic acid; C18:1, oleic acid; C18:2, linoleic acid; C18:3, gamma-linolenic acid.

Source: Data from Collaa et al. (2004); Australian Spirulina (2014).

Amino Acids (% DM)

Essential Amino Acids	
Isoleucine	4.13
Leucine	5.80
Lysine	4.00
Methionine	2.17
Phenylalanine	3.95
Threonine	4.17
Tryptophane	1.13
Valine	6.00

Continued

Nonessential Amino Acids	
Alanine	5.82
Arginine	5.98
Aspartic acid	6.34
Cystine	0.67
Glutamic acid	8.94
Glycine	3.50
Histidine	1.08
Proline	2.97
Serine	4.00
Tyrosine	4.60

Minerals (mg/kg DM)

Calcium	1315
Potassium	15,400
Zinc	39
Magnesium	1915
Manganese	25
Selenium	0.40[a]
Iron	580
Phosphorus	8942

[a] mg.

Source: Data from Australian Spirulina (2014).

Arthrospira maxima (Setchell & N.L. Gardner) Geitler 1932 (= *Spirulina maxima*)

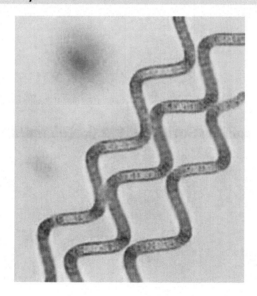

Order: Chroococcales

Family: Spirulinaceae

Common name: Spirulina.

Distribution: Africa, Asia, Australia, New Zealand, Europe (AlgaeBase).

Habitat: Lakes.

Description: This species is characterized by its regularly coiled trichomes. Under some conditions, its helical filaments may convert to abnormal morphologies, such as irregularly curved and even linear shapes. Filaments are usually 5–10 μm wide and 200–300 μm long. Surface of the filaments shows that the end cells of spiral trichomes are round or calyptrate, which are distinctive characteristics of this species. This species is a popular food supplement (Yeh et al., 2004; Wang and Zhao, 2005).

23

Nutritional Facts

Proximate Composition (% DM)

Protein	CHO[a]	Lipids
60–71	13–16	6–7

[a] Carbohydrate.

Amino Acids (g/100 g protein)

Isoleucine	6.0
Leucine	8.0
Valine	6.5
Lysine	4.6
Phenylalanine	4.9
Tyrosine	3.9
Methionine	1.4
Cysteine	0.4
Tryptophan	1.4
Threonine	4.6
Alanine	6.8
Arginine	6.5
Aspartic acid	8.6
Glutamic acid	12.6

Glycine	4.8
Histidine	1.8
Proline	3.9
Serine	4.2

Source: Data from Becker (2006).

Fatty Acids (% of lipids)

12:0	T[a]
14:0	0.3
14:1	0.1
15:0	T
16:0	45.1
16:1	6.8
16:2	T
17:0	0.2
18:0	1.4
18:1	1.9
18:2	14.6
18:3	20.3
20:3	0.8

[a] T, traces.

Source: Data from Becker (1994).

Phylloderma sacrum Suringer 1872 (= *Aphanothece sacrum*)

Order: Chroococcales

Family: Cyanobacteriaceae

Common name: Not available.

Distribution: Asia: Japan (AlgaeBase).

Habitat: Freshwater/Marine.

Description: It is a nonfilamentous cyanobacterial species. Colonies of this species are rigid and show a rather simple morphology: a bunch of individual cells, clearly separated from each other, are embedded in a globular gel-like matrix. This species has been used as a side dish by Japanese people from ancient times (John, 2002).

Nutritional Facts

Proximate Composition (g/100 g DM)

Protein	Fat	CHO[a]	Ash
26.85	0.11	64.3	7.5

[a] Carbohydrate.

Minerals (g/100 g DM)

Ca	P	Fe[a]
1.8	0.16	150

[a] mg%.

Source: Data from Johnston (1970).

Gloeocapsa livida (Carmichael) Kützing 1847 (= *Palmella livida*)

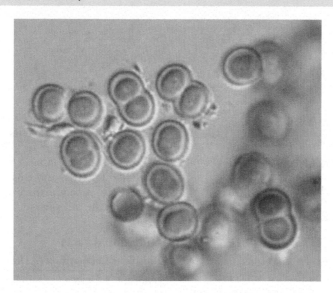

Order: Chroococcales

Family: Microcystaceae

Common name: Not available.

Distribution: Europe, Southwest Asia (AlgaeBase).

Habitat: Growing on limestone in freshwaters.

Description: It is a unicellular colonial species. Colonies are composed of groups of cells, which are closed in mucilaginous, usually wide, and concentrically lamellated envelopes. Cells are situated in colonies irregularly, more or less distant from one another, in old colonies in great numbers. Mucilaginous envelopes are fine, but layered and limited. Cells are spherical, only shortly after division hemispherical, pale blue-green or olive-green, usually with slightly granular content. This species may serve as a food supplement (CyanoDB.cz).

25

Nutritional Facts

Proximate Composition (% DM)

CHO[a]	Protein	Amino Acid	Lipid
18.0	1.8	1.7	12.5

[a] Carbohydrate.

Minerals (µg/ml)

Ca	Mn	Fe	Zn	Ni	Mg
29.0	344.4	4641	137.5	13.0	18,494

Source: Data from Rajeshwari and Rajashekhar (2011).

Synechococcus elongatus (Nägeli) Nägeli 1849

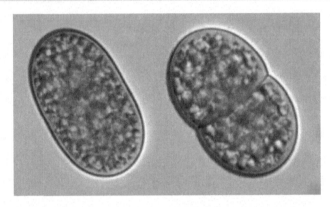

Order: Synechococcales

Family: Synechococcaceae

Common name: Not available.

Distribution: North America, Southwest Asia (Israel and Pakistan), Asia: China (AlgaeBase).

Habitat: Freshwater/terrestrial species.

Description: This species has a rod-shaped appearance, is unicellular, and may appear in the environment as isolated, paired, linearly connected, or in small clusters. It is a Gram-negative cell measuring 0.6 and 1.6 µm in size, with an inner and an outer cell membrane enveloping a cell wall. This cyanobacterium is able to swim or glide despite lacking flagella or cilia. It moves in a wave-like manner, and this movement is possibly accomplished via projections from the cell surface.

Nutritional Facts

Proximate Composition (% DM)

Protein	CHO[a]	Lipids
63	15	11

[a] Carbohydrate.

Source: Data from Becker (2006).

26

Lyngbya limnetica Lemmermann 1898 (= *Lyngbya lagerheimii*)

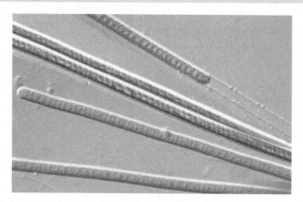

Order: Oscillatoriales

Family: Oscillatoriaceae

Commmon name: Mermaid's hair.

Distribution: Europe, North America, Southwest Asia, Australia and New Zealand (AlgaeBase).

Habitat: Freshwater habitats; rock mountain lake species.

Description: Filaments of this species are single or occasionally intertwined, coiled irregularly, or almost straight. Cells are 1.8–2.5 µm wide and 1.2–4 µm long. Cross-walls are not narrowed and are sometimes with a granule. End cell

is rounded, not tapering, and sheath is thin and colorless. This species is a food supplement (John, 2002).

Nutritional Facts

Proximate Composition (% DM)

CHO[a]	Protein	Amino Acid	Lipid
20.0	3.1	5.3	14.5

[a] Carbohydrate.

Minerals (µg/ml)

Ca	Mn	Fe	Zn	Ni	Mg
312.0	379.6	5456	134.1	2.2	18,650

Source: Data from Rajeshwari and Rajashekhar (2011).

Oscillatoria acuminata Gomont 1892 (= *Phormidium acuminatum*)

Order: Oscillatoriales

Family: Oscillatoriaceae

Common name: Not available.

Distribution: Europe, Asia, Australia and New Zealand (AlgaeBase).

Habitat: Hypersaline thermal springs; attached to hard substrate or soil; to 3 m deep in freshwater.

Description: Filaments of this species are solitary, usually straight or irregularly screw-like, coiled at ends, and their sheath is hyaline and coloress.

Individual cells measure 5 µm wide and 6 µm long. Apical cells are acute, conical, thorn-like or needle-like, without calyptras. No constriction at cross-walls or slightly constricted (Dadheech et al., 2013).

Oscillatoria calcuttensis Biswas 1925

Order: Oscillatoriales

Family: Oscillatoriaceae

Common name: Not available.

Distribution: India (Ramesh et al., 2013).

Habitat: Freshwater habitats.

Description: Thallus of this species is leathery brown and its trichomes, measuring 2 × 14 µm, are straight and parallel. End cells are conical, pointed, and not capitate (Ramesh et al., 2013; Deka and Sarma, 2011).

Oscillatoria foreaui Frémy 1942 (= *Phormidium froeaui*)

Order: Oscillatoriales

Family: Oscillatoriaceae

Common name: Not available.

Distribution: Southwest Asia: Punjab; Asia: Nepal (AlgaeBase).

Habitat: Freshwater habitats.

Description: This species forms long filaments of cells that can break into fragments called hormogonia. The hormogonia can grow into a new, longer filament. Breaks in the filament usually occur where dead cells (necridia) are present. Each unbranched filament (trichome) is made up of rows of cells measuring 1.3–2.0 µm diameter and 1.5–3.0 µm long. The tip of the trichome oscillates like a pendulum. This species serves as a good food supplement (Sharp, 1969).

Nutritional Facts

Proximate Composition (% DM)

	CHO[a]	Protein	Amino Acid	Lipid
O. acuminata	14.0	5.5	5.5	14.0
O. calcuttensis	9.6	2.2	2.1	20.0
O. foreaui	8.0	5.5	5.5	8.0

[a] Carbohydrate.

Minerals (µg/ml)

	Ca	Mn	Fe	Zn	Ni	Mg
O. acuminata	15.2	161.9	3365	53.0	5.4	21,050
O. calcuttensis	13.6	255.7	2238	80.0	5.8	18,600
O. foreaui	57.7	539.2	6402	211.2	9.4	12,812

Source: Data from Rajeshwari and Rajashekhar (2011).

PHYLUM: CHAROPHYTA (STONEWORT ALGAE)

Spirogyra sp.

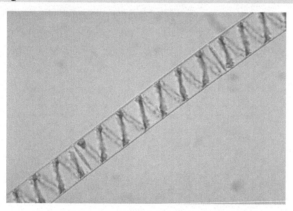

Phylum: Charophyta

Class: Zygnematophyceae

Order: Zygnematales

Family: Zygnemataceae

Distribution: Worldwide (Graphics and Web Programming Team, 2003).

Habitat: Quiet freshwater, streams, and lakes.

Description: Thalli are composed of unbranched uniseriate filaments intertwined to form skeins. Cells are cylindrical, 10 to >200 µm in diameter, most 20–60 µm, and up to several times as long. Cell wall is two-layered with inner cellulose, outer mucilage layer that makes filaments very slimy to touch. End walls are plane, and there are no flagellated stages.

Basal cells are infrequent with rhizoidal holdfasts. Cells are uninucleate and chloroplasts are from 1 to 15 per cell. Plastids are ribbon-like, coiled. Cytoplasmic strands support centrally located nucleus within central vacuole. The oil derived from this species may be more suitable to utilize as a food source (such as supplements) (Andini, 2009).

Nutritional Facts

Proximate Composition (% DM)

Protein	CHO[a]	Lipids
6–20	33–64	11–21

[a] Carbohydrate.
Source: Data from Becker (2006).

Fatty Acid Composition (% DM)

14:0	2.6
16:0	48.2
16:1	10.8
18:0	1.1
18:1	9.1
18:2	8.4
18:3	16.0
18:4	0.9
20:0	0.9
20:4	0.6
20:5	0.6
22:0	0.6
22:6	0.2

Source: Data from Iyanova et al. (2002).

PHYLUM: HAPTOPHYTA (COCCOLITHOPHORIDS)

Diacronema vlkianum Prauser 1958

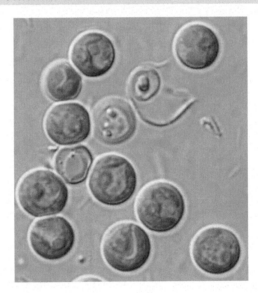

Phylum: Haptophyta

Class: Pavlovophyceae

Order: Pavlovales

Family: Pavlovaceae

Common name: Microalga.

Distribution: Europe: Britain (AlgaeBase).

Habitat: Freshwater/marine species.

Description: This species has both mobile and nonsymmetric cells. Cells of this species are orbiculate to heart

30

shaped in the dorsiventral view with a size of 3.5–8 × 3–6 µm and 1.5–3 µm thick. Anterior flagellum is 7 × 10 µm and posterior flagellum 3–9 µm. Haptonema is about 1 µm long. Hair points are seen at the tip of the flagella. Lateral hairs are absent. As this species synthesizes eicosapentaenoic (EPA) (a PUFA), it may be incorporated into the daily diet. Further, this species has applications in the preparation of food gels (John, 2002).

Nutritional Facts

Proximate Composition (% DM)

Protein	CHO[a]	Lipid
57.0	32.0	6

[a] Carbohydrate.

Source: Data from Gouveia et al. (2008).

PHYLUM: RHODOPHYTA (RED ALGAE)

Porphyridium cruentum (S.F. Gray) Nägeli 1849 (= *Porphyridium purpureum*)

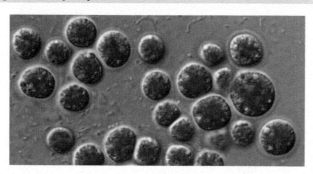

Phylum: Rhodophyta
Class: Porphyridiophyceae
Order: Porphyridiales
Family: Porphyridiaceae
Common name: Red alga.
Distribution: Europe, Asia, Australia and New Zealand (AlgaeBase).
Habitat; Freshwater/terrestrial species.
Description: This species has the tendency to form aggregated masses within mucilaginous cover leading to the formation of irregular colonies. The cells are globular in shape, and a distinctly star-shaped chloroplast with a pyrenoid is seen in the center. The diameter of the cells varies from 6 to 12 µm. This species has been reported to produce sulfated galactan exopolysaccharide, which could replace carrageenan in industrial applications (Gouveia et al., 2008).

Nutritional Facts

Proximate Composition (% DM)

Protein	CHO[a]	Lipids
28–39	40–57	9–14

[a] Carbohydrate.

Source: Data from Priyadarshani and Rath (2012).

Lemanea australis Atkinson 1890

Phylum: Rhodophyta
Class: Florideophyceae

Order: Batrachospermales
Family: Lemaneaceae

31

Common name: Not available.

Distribution: North America: Arkansas (AlgaeBase).

Habitat: Rivers, streams.

Description: The plants are homothallic, and both male and female reproductive organs are borne upon internal, peripheral, generative filaments. It is a diphasic organism whose haploid number of chromosomes is 10. The spermatium is liberated when its nucleus is in prophase, and the division is completed when the spermatium makes contact with the trichogyne. The spermatium appears to have no distinct wall at the time of its liberation, but develops one by the time it makes contact with the trichogyne. The usually bifurcated trichogyne of this species has no perceptible nucleus at any stage of development. No auxiliary cells occur in this plant, but there are many nutritive cells produced by the sterile laterals and the gonimoblast filaments (Mullahy, 1952).

Lemanea torulosa (Roth) C. Agardh 1814

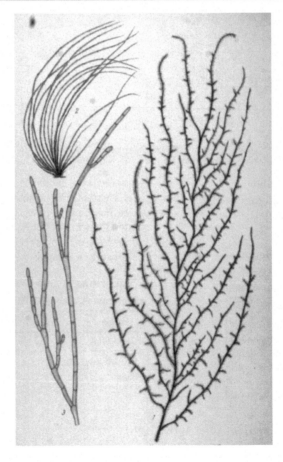

Order: Batrachospermales

Family: Lemaneaceae

Common name: Not available.

Distribution: Europe: Spain (Algae-Base).

Habitat: Attached to stones in flowing, freshwater habitats.

Description: Plants are often branched and look like tufts of stiff, dark brown, or black hair. Individual hairs are thickened at regular intervals forming nodes, the diameter of which is less than 365 μm. Spermatangial sori are arranged in patches and constricted rings. Each cell of the male filament bears a whorl of four small celled laterals and nine short shoots. It is sold as a foodstuff (Hoek, 1995).

Lemanea fluviatilis (Linnaeus) C. Agardh 1824

Order: Batrachospermales

Family: Lemaneaceae

Common name: Not available.

Distribution: Europe: Austria, Britain, Italy, Germany, Spain, Romania (Algae-Base).

Habitat: Running waters.

Description: This alga is greenish black (with a fleshy smell), thread-like, and branched, with a size ranging from 5 to 7 cm in length. Thalli of algae are narrowed abruptly toward the base, forming a thin cylindrical stalk. Branches are fasciculate, arcuate, long, pedicelled, and four in number. Axis and branches show distinct nodes. This species is reported to prefer noncalcareous substrates. It is sold as a foodstuff in India (Bhosale et al., 2012; Sophia et al., 2009; SBSAC).

Lemanea mamillosa Kützing 1845 (= *Lemanea fucina*)

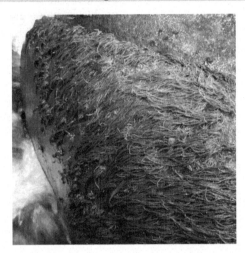

Order: Batrachospermales

Family: Lemaneaceae

Common name: Not available.

Distribution: Cosmopolitan in the northern hemisphere; Europe: Estonia, Finland, France, Germany, Sweden (AlgaeBase).

Habitat: Streams and rivers.

Description: This edible species largely resembles *L. fluviatilis* but has very long (up to 20 cm) and freely branched thalli (Kučera and Marvan, 2004).

Lemanea catenata (Kützing) (M.L. Vis & R.G. Sheath) (= *Paralemanea catenata*)

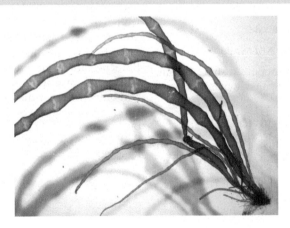

Order: Batrachospermales
Family: Lemaneaceae
Common name: Not available.
Distribution: Northern hemisphere; Europe: Belgium, France, Germany, Romania, and Spain; North America: California (AlgaeBase).
Habitat: Fast-flowing and turbulent waters.
Description: Plants are unstalked and unbranched, with multicellular thallus measuring a total length of 11.1 cm. It is an agarophyte producing hydrocolloid agar in its cell walls. It has tetrasporangia and a filamentous gonimoblast. Cells have multiple nuclei and plastids, but without flagella and centrioles. Chloroplasts lack external endoplasmic reticulum. Triphasic alternation of generations is seen in its life history (Marine Species Dictionary).

Nutritional Facts

Proximate Composition (% DM)

	L.a	L.t	L.f	L.m	L.c
Moisture	79.90	72.73	81.82	81.99	78.47
Ash	7.90	8.75	31.07	26.35	16.57
Fiber	1.09	2.79	3.03	2.58	2.63
Protein	25.80	23.80	31.07	17.48	24.44
Lipid	1.10	1.17	1.63	0.81	0.93
CHO[a]	38.20	27.20	20.51	27.40	48.60

L.a, *L. australis*; L.t, *L. torulosa*; L.f, *L. fluviatilis*; L.m, *L. mamillosa*; L.c, *L. catenata*.
[a] Carbohydrate.

Aminoacids (% DM), Carotenoid (mg/g FW) and Minerals (mg/100 g DM)

	L.a	L.t	L.f	L.m	L.c
Amino acids	12.80	15.60	17.20	9.60	13.60
Carotenoid	0.58	0.61	0.61	0.58	0.56
N2	111.48	113.79	226.28	214.17	134.39
P	150.70	142.88	135.17	72.42	72.92
K	836.92	804.30	1003.20	815.64	862.20
Na	141.90	146.80	423.32	276.10	238.09
Ca	111.44	112.23	113.76	111.59	111.90
Mg	70.87	99.85	135.51	135.65	120.20
Fe	16.95	19.07	26.18	25.64	18.17
Mn	5.06	6.51	14.34	8.03	9.22
Zn	3.93	5.25	3.77	1.86	1.22
Cu	12.98	13.08	8.91	12.13	13.09
Co	10.40	10.74	7.98	8.96	9.02

L.a, *L. australis*; L.t, *L. torulosa*; L.f, *L. fluviatilis*; L.m, *L. mamillosa*; L.c, *L. catenata*;
FW, fresh weight.
Source: Data from Singh and Gupta (2011).

Minerals of *L. australis* (mg/kg DM)

Mg	Al	Si	P	S	Cl	K	Ca
619	14,512	48,773	3067	28,193	3566	42,667	4175

Source: Data from Devi et al. (2011).

PHYLUM: EUGLENOZOA (FLAGELLATE PROTOZOA)

Euglena gracilis O.F. Müller 1786

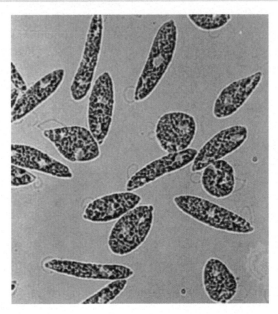

Phylum: Euglenozoa

Class: Euglenoidea

Order: Euglenales

Family: Euglenaceae

Common name: Not available.

Distribution: Common and ubiquitous all over Europe, western hemisphere, China, India (Agaebase).

Habitat: Primarily found in freshwater habitats, but they can also inhabit marine and soil environments.

Description: It is a two-flagellated microorganism with an exquisite spiral exoskeleton called a pellicle, as well as many other novel cell structures, such as nucleus, mitochondria, chloroplasts, Golgi bodies, lysosomes, vacuoles, an endoplasmic reticulum, an eyespot, and a few other eukaryotic organelles. Cell is approximately 50 μm long and 10 μm wide. This is a food supplement (MicrobeWiki, 2011).

Nutritional Facts

Proximate Composition (% DM)

Protein	CHO[a]	Lipids
39–61	14–18	14–20

[a] Carbohydrate.

Source: Data from Becker (2006).

36

PLANTS

PHYLUM: MAGNOLIOPHYTA (FLOWERING PLANTS)

Class: Magnoliopsida (Dicotyledons)

Justicia americana (Linnaeus)

Phylum: Magnoliophyta

Class: Magnoliopsida

Order: Lamiales

Family: Acanthaceae

Common name: American water-willow.

Distribution: Native to North America.

Habitat: On or near the shorelines of still or slow waters in lakes and rivers, and on rocky riffles and shoals in faster-flowing rivers.

Description: An herbaceous, aquatic flowering plant forming large colonies. They are partially submerged, reaching up to 40 cm tall from a creeping rhizome. Leaves are 10 cm opposite, sessile, linear or lanceolate, and slightly crenulated. Flowers are bicolored, borne in an opposite arrangement on spikes 3 cm long coming off a peduncle 10 cm long. Color ranges from white to pale lavender, with the upper corolla lip pale violet or white, arching over the lower lip mottled in dark purple. The lateral lobes are unadorned or slightly blushed. The root is edible.

Nutritional Facts

Proximate Composition (% DM)

Ash	Fat	Protein
8.19	9.37	47.4

Amino Acids (% DM)

Arginine	3.07
Histidine	1.08
Isoleucine	2.61
Leucine	4.38

Methionine	0.97
Phenylalanine	2.90
Threonine	2.26
Valine	2.60

Source: Data from Boyd (1968b).

Bacopa monniera (L.) (Pennell)

Order: Lamiales

Family: Plantaginaceae

Common name: Indian pennywort, water hyssop.

Distribution: Native to the wetlands of southern India; Australia, Europe, Africa, Asia, and North and South America; throughout India, Nepal, Sri Lanka, China, Pakistan, Taiwan, and Vietnam. It is also found in Florida, Hawaii, and other southern states of the United States.

Habitat: Damp conditions by a pond or bog garden.

Description: The leaves of this plant, which are edible, are succulent, oblong, and 4–6 mm thick. They are oblanceolate and are arranged oppositely on the stem. Flowers are small and white, with four or five petals. It can even grow in slightly brackish conditions. Propagation is often achieved through cuttings.

Nutritional Facts

Proximate Composition of Leaves (g/100 g FW)

Moisture	Protein	Fat	CHO[a]	Fiber	Ash	Energy[b]
88.4	2.1	0.6	5.0	1.05	1.9	38

FW, fresh weight.
[a] Carbohydrate.
[b] Calories.

Minerals and Vitamins (mg/100 g)

Ca	Fe	P	Ascorbic Acid	Nicotinic Acid
202	7.8	16	63	0.3

Source: Data from Sabinsa Corporation (2001); Zhou et al. (2007).

38

Mentha arvensis (Linnaeus)

Order: Lamiales

Family: Lamiaceae

Common name: Wild mint, corn mint.

Distribution: Native to the temperate regions of Europe and western and central Asia; east to the Himalayas and eastern Siberia, and North America.

Habitat: Higher elevations along creeks and in meadows.

Description: It is an herbaceous perennial plant generally growing to 10–60 cm and rarely up to 100 cm tall. It has a creeping rootstock from which grow erect or semisprawling squarish stems. Leaves are in opposite pairs, simple, 2–6.5 cm long and 1–2 cm broad, hairy, and with a coarsely serrated margin. Flowers are pale purple (occasionally white or pink), in whorls on the stem at the bases of the leaves. Each flower is 3–4 mm (0.12–0.16 in.) long and has a five-lobed hairy calyx, a four-lobed corolla with the uppermost lobe larger than the others, and four stamens. Its leaves are edible.

Nutritional Facts

Proximate Composition (g/100 g)

Moisture	Protein	Starch	Lipid	Fiber	Ash	Energy[a]
88.1	3.2	0.8	2.0	2.5	1.5	147

[a] kJ.

Minerals (mg/100 g)

Ca	Mg	Fe	K	Cu
281	37	124	420	1.6

Amino Acids (g/100 g total amino acids)

Aspartic acid	12.6
Threonine	5.1
Serine	4.8
Glutamic acid	15.5
Proline	6.0
Glycine	5.4
Alanine	6.5
Valine	4.9
Methionine	2.0
Isoleucine	3.4
Leucine	9.1
Tyrosine	4.6
Phenylalanine	6.1
Histidine	2.2
Lysine	6.3
Tryptophan	1.0
Arginine	4.4

Source: Data from Yeoh and Wong (1993).

39

Nymphaea odorata (Aiton) (= *Nymphaea tuberosa*)

Order: Nymphaeales

Family: Nymphaeaceae

Common name: Fragrant water lily, American water lily, white water lily.

Distribution: Throughout North America, where it ranges from Central America to northern Canada.

Habitat: Shallow lakes, ponds, and permanent slow-moving waters.

Description: The plant is rooted from a branched rhizome, which gives rise to long petioles that terminate in smooth floating leaves. Leaves are round with a waxy upper coating that is water repellent. Flowers also float. They are radially symmetric with prominent yellow stamens and many white petals. The flowers open each day and close again each night and are very fragrant. The seeds, rhizomes, flowers, and leaves of this plant are edible.

Nutritional Facts

Proximate Composition (% DM)

Ash	Fat	Protein
4.24	8.57	40.0

Amino Acids (% DM)

Arginine	2.83
Histidine	1.07
Isoleucine	1.97
Leucine	3.82
Methionine	0.71
Phenylalanine	2.24
Threonine	1.91
Valine	2.62

Source: Data from Boyd (1968b).

Minerals (% or mg/100 g DM)

Ca (%)	P (%)	K (%)	Na (%)	Mg (%)	Fe (mg)	Mn (mg)	Mo (mg)
2.76	0.13	1.56	0.165	0.440	391	112	14.4

Source: Data from Linn et al. (1975).

Nuphar variegata (Engelm. ex Durand)

Order: Nymphaeales

Family: Nymphaeaceae

Common name: Variegated pond lily, yellow pond lily.

Distribution: Native to much of Canada and the northernmost United States.

Habitat: Lakes or ponds, rivers, or streams.

Description: Leaves of this species are oval to heart shaped, 10–25 cm long, and about two-thirds as wide, with a rounded tip and deep cleft at the base; the lobes at the base are rounded and about half as long as the rest of the blade. Leaf stalks are flattened on one side, and are narrowly "winged." Leaves are flat and float on the surface. A single flower of 2.5–5 cm across is seen on a stout stalk rising above the water. Flowers are globular, typically with three round bright yellow petals that may be green on the outer surface near the base, and dark red on the inside near the base, with three smaller green sepals separating the petals. Roots of this species are edible (Minnesota Wildflowers.info).

Nutritional Facts

Proximate Composition (% DM)

Protein	Ash	Fiber
15.70	0.96	23.13

Minerals (% or mg/100 g DM)

Ca (%)	P (%)	K (%)	Na (%)	Mg (%)	Fe (mg)	Mn (mg)	Mo (mg)
0.52	0.23	1.62	0.508	0.151	583	115	10.2

Source: Data from Linn et al. (1975).

Nelumbo lutea (Willd.)

Order: Proteales

Family: Nelumbonaceae

Common name: American lotus, yellow lotus.

Distribution: Native to North America; southeastern United States, Mexico, Honduras, and the Caribbean.

Habitat: Marshes, quiet backwaters, and nearshore areas in large rivers; muddy, shallow waters to water over 6 ft deep.

Description: It is a floating-leaf aquatic plant that often rises above the surface of the water. It has round, bluish green leaves that can be up to 60 cm in diameter and that are flat in appearance if the plant is floating and conical when emergent. Flowers are very large (up to 25 cm across) and may have more than 20 delicate petals, which range in color from yellowish white to darker yellow. Seed of this species is edible (Rhode Island Department of Environmental Management, 2010).

Nutritional Facts

Proximate Composition (% DM)

Ash	Protein	Fat	Cellulose	Energy[a]
10.3	13.7	5.3	23.6	3.7

[a] kcal/g.

Source: Data from Boyd (1968b).

Nelumbo nucifera (Gaertn.)

Order: Proteales

Family: Nelumbonaceae

Common name: Indian lotus, sacred lotus.

Habitat: Small ponds and shallow areas of lakes and rivers.

Distribution: Asia (Iran to Japan); northern Australia (Missouri Botanical Garden).

Description: It is a large-flowered lotus that typically grows 1–1.8 m tall in shallow water and spreads by thickened rhizomes rooted in the mud. This is a marginal aquatic perennial plant with rounded, parasol-like, upward-cupped, waxy green leaves (to 60 cm across) and long petioles. Large, cupped, fragrant, pink or white flowers (20–30 cm diameter) appear on stiff stems above the foliage. Rhizomes, leaves, and seeds of this species are edible and are sometimes used in Asian cooking (Missouri Botanical Garden).

Nutritional Facts

Proximate Composition of Seeds (% DM)

Ash	Moisture	Fat	Protein	CHO[a]	Fiber	Energy[b]
4.50	10.50	1.93	10.60	72.17	2.70	348.45

[a] Carbohydrate.
[b] Calories/100 g.

Minerals of Seeds (% DM)

Cr	Na	K	Ca	Mg	Cu	Zn	Mn	Fe
0.004	1.00	28.5	22.10	9.20	0.046	0.084	0.356	0.199

Source: Data from Indrayan et al. (2005).

Nutritional Facts of Rhizome (g/100 g Flour)

Total ash	1.10
Total nitrogen	1.36
Total protein	8.48
Water-soluble proteins	1.23
Salt-soluble proteins	5.73
Total salt-soluble proteins	6.064
Reducing sugars	0.168
Nonreducing sugars	18.87
Total sugars	19.08
Free amino acid	0.78

Source: Data from Shad et al. (2011).

Fats and Fatty Acids of Root (per 120 g)

Total fat	84 mg
Saturated fat	25 mg
Monounsaturated fat	17 mg
Polyunsaturated fat	17 mg
Omega-3 fatty acids	4.8 mg
Omega-6 fatty acids	12 mg

Vitamins of Root (per 120 g)

Vitamin C	33 mg
Vitamin E	12 µg

Vitamin K	0.12 µg
Thiamin	152 µg
Riboflavin	12 µg
Niacin	360 µg
Vitamin B6	262 µg
Folate	9.6 µg
Pantothenic acid	362 µg
Choline	30.5 mg

Source: Data from Nutrition Facts (2014).

Alternanthera sessilis (L.) (R.Br. ex DC)

Order: Caryophyllales

Family: Amaranthaceae

Common name: Sessile joyweed, dwarf copperleaf.

Distribution: Tropical regions of the world, especially tropical America, Africa, and Asia.

Habitat: Ditches, swamps, gardens, rice fields.

Description: It is a perennial herb growing annually up to 1 m tall, erect, ascending or creeping, often widely branched, with strong taproot. Leaves obovate to broadly elliptic, occasionally linear-lanceolate, 1–15 cm long, 0.3–3 cm wide, glabrous to sparsely villous, petioles 1–5 mm long. Flowers are in sessile spikes, and bract and bracteoles are shiny white, 0.7–1.5 mm long, glabrous. Sepals are equal, 2.5–3 mm long. Its leaves and young shoots are eaten as a vegetable, particularly in tropical Africa, or cooked in soup.

Nutritional Facts

Proximate Composition (% DM)

Protein	Fat	Fiber	Moisture	Ash	CHO[a]
17.60	5.61	16.41	9.26	12.44	38.67

[a] Carbohydrate.

Minerals (mg/100 g)

Na	Ca	Fe	P	K	Zn	Cu	Mg	Mn
646	2770	0.78	94.10	668.50	9.81	1.77	1195	19.30

Source: Data from Gbadamosi and Okolosi (2013).

Polygonum barbatum (L.)

Order: Caryophyllales

Family: Polygonaceae

Common name: Jointweed, knotgrass, smartweed.

Distribution: Southeast Asia, Malaysia to tropical Australia; tropical Africa.

Habitat: In shallow fresh or brackish water; pools, marshy sites, along drains; tidal rice fields.

Description: Terrestrial or aquatic, perennial, erect herb, up to 2 m long. Roots are fibrous, white, or brown. Stems are erect or procumbent, round, hollow, thickened at nodes, glabrous, or hair. Leaves are simple, not lobed or divided; alternate spiral, stalked, more than 2 cm long/wide, hairy on both sides, margin entire, apex acute, base acute, or rounded. Flowers are bisexual, grouped together into a terminal spike, stalked, white or pink, petals not visible. Shoot is edible in this species (http://www.oswaldasia.org/species/p/polba/polba_en.html).

Nutritional Facts (shoot, % FW)

Lignin	Fat	CHO[a]	Protein	P	N2	K	Na
10	2	3.7	7.5	0.44	1.20	0.015	0.035

[a] Carbohydrate.

Source: Data from Jain et al. (2011).

45

Rumex nepalensis (Sprengel)

Order: Caryophyllales

Family: Polygonaceae

Common name: Nepal dock.

Distribution: Afghanistan, Bhutan, India, Indonesia, Japan (introduced), Myanmar, Nepal, Pakistan, Sikkim, Tajikistan, Vietnam; Southwest Asia.

Habitat: Grassy slopes, moist valleys, along ditches.

Description: This plant is erect with long taproots. Stems are erect, 50–100 cm tall, branched above, glabrous, grooved. The fleshy to leathery leaves form a basal rosette at the root. Leaf blade is broadly ovate, 10–15 × 4–8 cm. The basal leaves may be different from those near the inflorescence. The inconspicuous flowers are carried above the leaves in whorl-like clusters. They are mostly hermaphrodite, or they may be functionally male or female. The shoot is edible in this species (www.eFloras.org).

Nutritional Facts

Proximate Composition (% FW)

Plant Part	Ash	Moisture	Protein	Fats	Fiber	CHO[a]
Root	6.40	4.79	11.12	15.57	13.50	48.62
Stem	9.73	5.30	15.30	18.69	10.78	40.20
Leaf	8.11	3.50	13.95	17.54	15.38	41.52
Flower	8.77	4.82	9.88	19.10	9.00	48.43
Fruit	4.95	3.00	14.84	11.78	14.60	50.83
Seed	5.47	3.40	18.53	13.80	17.40	41.40

[a] Carbohydrate.

Source: Data from Hameed and Dastagir (2009).

Rumex hastatus (D. Don)

Order: Caryophyllales

Family: Polygonaceae

Common name: Arrowleaf dock, yellow sock, curled sock.

Distribution: Northern hemisphere; northern Pakistan, northeast Afghanistan, southwest China.

Habitat: Wetlands.

Description: It is a bushy and richly branched shrub with many ascending stems 90–120 cm tall. Stems are woody at base. Leaves are narrow and arrow shaped with a pair of narrow spreading basal lobes. Stems have numerous thin branches with terminal very slender clusters of distant whorls of tiny greenish pink or pinkish green flowers. Flowers are very small and hermaphrodite. Tender young leaves and shoots are eaten raw or cooked (http://www.flowersofindia.net/catalog/slides/Arrowleaf%20Dock.html).

Nutritional Facts

Proximate Composition (% FW)

Plant Part	Ash	Moisture	Protein	Fats	Fiber	CHO[a]
Root	5.25	20.22	13.65	2.54	19.37	39.04
Stem	6.23	26.80	6.05	4.50	19.26	36.19
Leaf	18.58	22.56	14.00	5.84	14.57	24.50
Flower	8.66	20.50	9.50	3.54	18.65	39.24
Fruit	9.73	24.01	13.59	4.57	15.06	33.04
Seed	15.50	23.05	12.01	4.56	16.06	28.88

[a] Carbohydrate.

Source: Data from Hameed and Dastagir (2009).

Rumex dentatus (Linn.)

Order: Caryophyllales

Family: Polygonaceae

Common name: Toothed dock.

Distribution: Native to parts of Eurasia and North Africa, Africa and Madagascar, Asia.

Habitat: Moist areas, such as lakeshores and the edges of cultivated fields.

Description: It is an annual or biennial herb producing a slender, erect stem up to 70 or 80 cm in maximum height. Leaves are lance shaped to oval with slightly wavy edges, growing to a maximum length around 12 cm. Inflorescence is an interrupted series of clusters of flowers, with 10–20 flowers per cluster. Leaves are edible in this species.

Nutritional Facts

Proximate Composition (% FW)

Plant Part	Ash	Moisture	Protein	Fats	Fiber	CHO[a]
Root	8.37	9.33	10.58	14.66	10.65	46.40
Stem	7.59	6.23	15.72	11.44	8.76	50.25
Leaf	8.65	4.05	13.75	12.50	9.03	52.05
Flower	7.50	5.86	14.76	13.00	10.88	40.00
Fruit	6.99	6.23	10.50	11.80	11.67	52.84
Seed	8.44	4.91	12.12	11.12	9.00	54.40

[a] Carbohydrate.

Source: Data from Hameed and Dastagir (2009).

Fagopyrum tataricum (L.) (Gaertn.)

Order: Caryophyllales

Family: Polygonaceae

Common name: Tartary buckwheat.

Distribution: Domesticated in East Asia, Southwest China, India.

Habitat: Wetlands, vegetable gardens.

Description: This is an annual growing to 0.8 m. It has five-petaled flowers, monoecious (individual flowers are either male or female, but both sexes can be found on the same plant), arranged in a compound raceme that produces laterally flowered cymose clusters. Flowers (2 cm diameter) are greenish or red in color and are actinomorphic in symmetry. Leaves are pinnately veined and have three sepals. Young leaves, stems, and blossoms of this species have been used for both medicinal and culinary purposes in Europe and Asia (Healthwithfood.org; Plants for a Future).

Fagopyrum esculentum (Moench)

Order: Caryophyllales

Family: Polygonaceae

Common name: Buckwheat.

Distribution: From south of the Himalayas to Bhutan, Nepal, northern India, and northern Pakistan; China, Korea, and Japan (Ratan and Kothiyal, 2011).

Habitat: Wetlands.

Description: It is a slender, annual herb 90–150 cm tall. Stems are red-tinged, slender, and leaves are broadly triangular and acute up to 7 cm long. This species has five-petaled white to pink flowers arranged in a compound raceme that produces laterally flowered cymose clusters. Young leaves, stems, and blossoms of this species have been used for both medicinal and culinary purposes in Europe and Asia (Healthwithfood.org; Ratan and Kothiyal, 2011).

Nutritional Facts

Chemical Elements (% DM)

F. tartaricum				
Plant Parts	**N2**	**P2O5**	**K2O**	**Ash**
Stalk, branches	1.71	0.365	2.94	10.7
Leaves	1.92	0.541	2.62	13.1
Roots	1.83	0.355	1.47	23.4
Fruits	1.72	0.793	1.26	2.39

F. esculentum				
Plant Parts	**N2**	**P2O5**	**K2O**	**Ash**
Stalk, branches	1.75	0.446	3.89	15.8
Leaves	2.22	0.541	3.21	14.1
Roots	1.68	0.384	1.58	27.5
Fruits	2.09	1.06	1.29	2.88

Amino Acids of Leaves (% to Protein of Hydrolysate, DM)

	F. tartaricum	*F. esculentum*
Asparaginic acid	9.4	10.8
Threonine	4.7	5.1
Serine	4.3	5.0
Glutamine acid	10.9	13.7
Proline	2.0	3.5
Glycine	5.5	5.8
Alanine	6.0	6.3
Valine	5.9	5.5
Methionine	0.91	1.3
Isoleucine	5.0	4.3
Leucine	12.0	9.6
Thyrosinum	4.8	4.9
Phenylalanine	7.7	6.2
Histidine	4.1	3.4
Lysine	6.2	5.8
Arginine	7.4	6.1
Sum of amino acids (mg on 100 g DM)	15,458	22,994

Source: Data from Lahanov et al. (2004).

Enhydra fluctuans (Lour.)

Order: Asterales

Family: Asteraceae

Common name: Kankong-kalabau.

Distribution: Tropical Africa and Asia to Malaya (Philippine Medicinal Plants).

Habitat: Ponds, swamps, and streams.

Description: It is a trailing, annual marsh herb, also floating on water. Stem is 30–60 cm long, rooting at the nodes. Leaves are sessile, 2.5–7.5 cm long, linear-oblong, acute, or obtuse, entire or subcrenate. Flowering heads are without stalks, borne singly in the axils of the leaves, and flowers are white or greenish white. Leaves and young shoots in this species are eaten as leafy vegetable (MeD India; Philippine Medicinal Plants).

Nutritional Facts

Proximate Composition (per DM)

Protein (%)	Ash (%)	Beta-carotene[a] (mg/100 g)
32.6	14.2	3.7

[a] Wet weight.

Minerals (% DM)

Na	K	Ca	P
2.6	2.3	0.7	0.7

Source: Data from Devanji et al. (1993).

Trapa natans (Linnaeus)

Order: Myrtales

Family: Lythraceae

Common name: Water caltrop, water chestnut, buffalo nut, bat nut, devil pod, ling nut.

Distribution: Native to warm temperate parts of Eurasia and Africa.

Habitat: Stagnant waters, lakes, channels with weak currents, ponds, and marshes.

Description: The submerged stem of this species reaches 3.6–4.5 m in length, anchored into the mud by very fine roots. It has two types of leaves: finely divided feather-like submerged leaves borne along the length of the stem, and undivided floating leaves borne in a rosette at the water's surface. Floating leaves have sawtooth edges and are ovoid or triangular in shape, 2–3 cm long. Four-petaled white flowers are seen. The nut-like fruit and seed of this species are edible.

Nutritional Facts

Proximate Composition (% DM)

Moisture	Protein	Lipid	Ash	CHO[a]	Fiber	Energy[b]
7.6	11.4	8.0	13.3	67.3	4.2	347

[a] Carbohydrate.
[b] kcal/100 g.

Minerals (mg/100 g)

Cu	Fe	Mn	Zn
1.84	72.00	11.00	8.20

Source: Data from Lim (2013).

Proximate Composition (% WW)

Moisture	Lipid	Ash	Fiber	Protein
81.12	0.36	1.33	0.72	1.87

Source: Data from Shalab et al. (2012).

Ceratophyllum demersum (Linnaeus)

Order: Ceratophyllales

Family: Ceratophyllaceae

Common name: Coontail.

Distribution: North America (United States and Canada except Newfoundland); Europe (Norway); China, Siberia, Burkina Faso (Africa), Vietnam, and New Zealand.

Habitat: Stagnant and slow-moving water; ponds, lakes, streams.

Description: It is a true rootless aquatic that floats just under the surface of the water. It develops rhizoids if anchored in the substrate. Stems reach lengths of 1–3 m, with numerous side shoots making a single specimen appear as a large, bushy mass. Leaves that are fan shaped and feathery are produced in whorls of 6–12, each leaf 8–40 mm long, simple, or forked into two to eight thread-like segments edged with spiny teeth; they are stiff and brittle. It is monoecious, with separate male and female flowers produced on the same plant. Flowers are small, 2 mm long, with eight or more greenish brown petals; they are produced in the leaf axils. Leaves of this species are edible.

Nutritional Facts

Proximate Composition (% DM)

Protein	Ash	Fiber
17.00	2.18	15.20

Minerals (% DM)

Ca	P	K	Na	Mg
2.46	0.22	1.29	0.336	0.686

Minerals (mg/100 g)

Fe	Mn	Mo
546	281	18.9

Source: Data from Little (1979).

53

Piper sarmentosum (Roxb.)

Order: Piperales

Family: Piperaceae

Common name: Wild pepper, betel leaves.

Distribution: Tropical areas of Southeast Asia, Northeast India, and South China, and as far as the Andaman islands.

Habitat: Riverbanks, on damp soil.

Description: It is an erect herb with long creeping stems, growing to more than 10 m. It is mostly creeping along the ground, and most parts are very finely powdery pubescent at least when young, dioecious. Fertile stems are erect. Leaf blades toward the base of the stem are ovate to suborbicular, and those toward the apex of the stem are four smaller, ovate or ovate-lanceolate, 7–14 × 6–13 cm. Flowers are bisexual or unisexual, in terminal or leaf opposite spikes. Male spikes are white, 1.5–2.5 cm × 2–3 mm. Leaves of this species are edible (Asia Herbs).

Nutritional Facts

Proximate Composition (g/100 g WW)

Moisture	Protein	Starch	Lipid	Fiber	Ash	Energy[a]
81.1	4.0	1.3	5.7	1.9	2.3	321

[a] kJ.

Minerals (mg/100 g)

Ca	Mg	Fe	K	Cu
396	80	187	572	3.8

Amino Acids (g/100 g total amino acids)

Aspartic acid	11.7
Threonine	5.0
Serine	5.7
Glutamic acid	14.4
Proline	6.3
Glycine	6.2
Alanine	6.5
Valine	4.7
Methionine	2.0
Isoleucine	3.3
Leucine	9.2
Tyrosine	5.4
Phenylalanine	6.0
Histidine	2.2
Lysine	6.5
Tryptophan	0.5
Arginine	4.4

Source: Data from Yeoh and Wong (1993).

54

Cardamine hirsuta (L.)

Order: Brassicales

Family: Brassicaceae

Common name: Hairy bittercress.

Distribution: Native to Europe and Asia, but also present in North America.

Habitat: Wetlands, damp, recently disturbed soil.

Description: This species grows to no more than 30 cm. It is an annual plant. Stems are hairless and the leaves do not clasp the stems. Each leaf generally contains four to eight leaflets arranged alternately along the rachis. Flowers occur in racemes. Each flower has four white petals, generally 3–5 mm in diameter. Leaves and flowers of this species are eaten raw or cooked.

Nutritional Facts

Proximate Composition of Shoot (% DM)

Fat	CHO[a]	Protein
3.0	8.0	14.4

[a] Carbohydrate.

Minerals (% DM)

P	N2	K	Na
0.66	0.66	0.44	0.016

Source: Data from Jain et al. (2011).

Nasturtium officinale (W.T. Aiton)

Order: Brassicales

Family: Brassicaceae

Common name: Watercress.

Distribution: Europe, including Britain, from Sweden and Denmark south and east to North Africa and West Asia (Plants for a Future, 1996–2012).

Habitat: Lakes, ponds, and in slow-moving water in rivers, canals, and streams; brooks, ditches, and pond margins.

Description: It is an emergent perennial herb with small floating leaves that are rounded, dark green, and waxy. Leaves are between 4 and 12 cm long, with the end leaflet typically being the largest. Its branching stems spread out for 2–3 cm over the surface. Slender roots hang down from the nodes of the stems. At the top of stems and short stalks, its flowers are 3–5 mm long and have four white petals. It is cultivated in several countries for its leaves and seeds. As a green vegetable its leaves are used as garnish and in salad (Plants for a Future, 1996–2012; D'Agaro, 2006).

Nutritional Facts

Proximate Composition (% DM)

Protein	EE[a]	Fiber	Ash	NFE[b]	Energy[c]
20.7	1.9	16.5	20.0	40.8	13.9

[a] Ether extract.
[b] N-free extract.
[c] MJ/kg.
Source: Data from D'Agaro (2006).

Proximate Composition of Leaves (g/100 g WW)

Protein	Water	CHO[a]	Fat	Fiber	Ash	Energy[a]
2.2	93.3	3.0	0.3	0.7	1.2	19

[a] Carbohydrate.
[b] Calories/100 g.

Minerals (mg per 100 g WW)

Ca	P	Fe	Mg	Na	K	Zn
151	54	1.7	0	52	282	0

Vitamins (mg per 100 g WW)

Vitamin A	Thiamin	Riboflavin	Niacin	B6	Vitamin C
2940	0.08	0.16	0.9	0	79

Source: Data from Plants for a Future (1996–2012).

Centella asiatica (L.) (Urban)

Order: Umbelliferae

Family: Apiaceae

Common name: Asiatic pennywort, Indian pennywort.

Distribution: Southeast Asia, India, Sri Lanka, China, South Africa, Southeast United States, eastern South America (Singh et al., 2010).

Habitat: Wetlands, swampy areas, ditches.

Description: The stems of this species are slender with creeping stolons. They are green to reddish green in color. It has long-stalked, green, reniform leaves with rounded apices that have smooth texture with palmately netted veins. The leaves are borne on pericladial petioles, around 2 cm. The flowers are white or pinkish to red in color, borne in small, rounded bunches (umbels) near the surface of the soil. Each flower is partly enclosed in two green bracts. The hermaphrodite flowers are minute in size (less than 3 mm), with five or six corolla lobes per flower. The whole plant of this species is edible.

Nutritional Facts

Proximate Composition of Shoot (% DM)

Fat	CHO[a]	Protein
1.0	7.0	8.25

[a] Carbohydrate.

Minerals (% or mg/100 g DM)

P (%)	N2 (%)	K (%)	Na (%)	Fe (%)	Mg (mg)	Cu (mg)	Zn (mg)
0.62	0.62	0.33	0.08	0.85	0.72	0.12	1.24

Source: Data from Jain et al. (2011).

Caltha palustris (Linnaeus)

Order: Ranunculales

Family: Ranunculaceae

Common name: Kingcup, marsh marigold.

Distribution: Temperate regions of the northern hemisphere, North America.

Habitat: Marshes, swamps, wet meadows, and stream margin; moist soil.

Description: It is an herbaceous perennial species growing up to 80 cm tall on hollow, branching stems. Upper stem leaves are smaller and stalkless. Leaves are rounded to heart shaped and are 3–20 cm across, with a bluntly serrated margin and a thick, waxy texture. Bright, shiny, five- or six-part yellow flowers (2.5–5.0 cm diameter) have five to nine waxy deep yellow petal-like sepals. All parts of this species are edible if boiled (Missouri Botanical Garden).

Nutritional Facts

Proximate Composition (% DM)

Protein	Ash	Fiber
13.93	1.28	17.40

Minerals (% or mg/100 g DM)

Ca (%)	P (%)	K (%)	Na (%)	Mg (%)	Fe (mg)	Mn (mg)	Mo (mg)
1.02	0.60	3.19	0.128	0.298	1231	498	23.8

Source: Data from Linn et al. (1975).

58

Hedyotis auricularia (Linnaeus)

Order: Rubiales

Family: Rubiaceae

Common name: Not available.

Distribution: Native to tropical and subtropical Asia and to islands of the Northwest Pacific.

Habitat: Wet grassland, streamsides.

Description: It is a widespread perennial herb 1 m tall. Stems are flattened, four-angled, sometimes glabrescent. Leaves of this species are subsessile to petiolate, 0.4–3 cm wide. Cyme is densely arranged, being head shaped, axillary, without common peduncle. Corolla is white, tubular, and its lobes measure 0.5–1 mm across. The shoot of this species is edible (Tao and Taylor, 2011).

Nutritional Facts

Proximate Composition of Shoot (% DM)

Fat	CHO[a]	Protein
3.0	7.3	7.88

[a] Carbohydrate.

Minerals (% DM)

P	N2	K	Na
0.4	0.4	0.12	0.01

Source: Data from Jain et al. (2011).

59

Oxalis corniculata (L.)

Order: Oxalidales

Family: Oxalidaceae

Common name: Creeping wood sorrel, procumbent yellow sorrel, sleeping beauty.

Distribution: Cosmopolitan; Asia: Afghanistan, Bhutan, Cambodia, China, Taiwan, India.

Habitat: Wetlands, cultivated beds.

Description: It is a somewhat delicate-appearing, low-growing, herbaceous plant with a narrow, creeping stem that readily roots at the nodes. Trifoliate leaves are subdivided into three rounded leaflets and resemble a cloverin shape. Some varieties have green leaves, while others have purple. Leaves have inconspicuous stipules at the base of each petiole. Flowers are hermaphrodite. The whole plant of this species is edible.

Nutritional Facts

Proximate Composition of Shoot (% DM)

Fat	CHO[a]	Protein
2.0	11.8	9.2

[a] Carbohydrate.

Minerals (% or mg/100 g DM)

P (%)	N2 (%)	K (%)	Na (%)	Fe (%)	Mg (mg)	Cu (mg)	Zn (mg)
0.92	1.47	0.30	0.02	0.90	1.85	0.15	1.13

Source: Data from Jain et al. (2011).

60

Viola pilosa (Blume)

Order: Violales

Family: Violaceae

Common name: Smooth-leaf white violet.

Distribution: Himalayas, from Afghanistan to Southwest China, Burma, and Southeast Asia; West Guangxi, Guizhou, Sichuan, Southeast Xizang, Yunnan, Afghanistan, Bhutan, India, Indonesia, Kashmir, Malaysia, Myanmar, Nepal, Sri Lanka, Thailand (www.eFloras.org).

Habitat: Wetlands, grasslands.

Description: It is a perennial herb with a very short stem. Stolon is elongated, slender, glabrous, with evenly scattered leaves. Leaves are narrower, longer, ovate-lance shaped, pointed to long-pointed, with thin white hairs or almost hairless above. Flowers are lilac to almost white, 1–1.5 cm across. Upper petals normally have hairs at the base. Sepals are lanceolate, 6–7.5 × ca. 2.5 mm, apex acute. Shoot of this species is edible (www.eFloras.org).

Nutritional Facts

Proximate Composition of Shoot (% DM)

Fat	CHO[a]	Protein
3.0	7.5	4.9

[a] Carbohydrate.

Minerals (% or µg/100 g DM)

P (%)	N2 (%)	K (%)	Na (%)	Fe (µg)	Mg (µg)	Cu (µg)	Zn (µg)
0.43	0.78	0.35	0.02	0.66	3.35	0.04	1.26

Source: Data from Jain et al. (2011).

Jussiaea repens (Linnaeus)

Order: Myrtales

Family: Onagraceae

Common name: Primrose willow.

Distribution: Native to United States, Philippines, Kenya, Australia (http://www.missouriplants.com/Yellowalt/Jussiaea_repens_page.html).

Habitat: Wetlands, shallow still water, muddy soil, pond margins.

Description: Stem of this species is erect to repent. It may be on land or floating, herbaceous, glabrous, or sparsely pubescent, often reddish, from fibrous roots, to about 50 cm long. Leaves are alternate, petiolate, glabrous, oblong to elliptic, tapering to base, acute at apex, entire, to 12 cm long and 2 cm broad. Inflorescence is single axillary flowers on long peduncles. Floral tube measures 1.2 cm in flower, five-angled. Petals are five in number, free, yellow, 1.7 cm long and 1.2 cm broad, and glabrous. Shoot of this plant is edible (http://www.missouriplants.com/Yellowalt/Jussiaea_repens_page.html).

Nutritional Facts

Proximate Composition of Shoot (% DM)

Fat	CHO[a]	Protein
1.0	11.3	13.5

[a] Carbohydrate.

Minerals (% or mg/100 g DM)

P (%)	N2 (%)	K (%)	Na (%)	Fe (%)	Mg (mg)	Cu (mg)	Zn (mg)
0.96	2.16	0.26	0.02	0.90	2.77	0.16	1.13

Source: Data from Jain et al. (2011).

Jussiaea suffruticosa (Linn.) (= *Ludwigia octovalvis*)

Order: Myrtales

Family: Onagraceae

Common name: Willow primrose.

Habitat: Wetlands, seasonally wet places, rice-growing areas.

Distribution: Throughout the tropics of the world (Kew Royal Botanic Garden).

Description: It is an erect, stout, well-branched robust herb. It may be woody at the base and shrubby at times, growing up to 4 m. Stems may be red-brown. Alternately arranged leaves are light green, narrowly lance shaped to ovate, up to 15 cm long, 0.4–4 cm wide, densely velvety on both sides, narrowed at base and tip. Sepals are four, ovate or lance shaped, 0.8–1.3 cm long by 1–7.5 mm wide. Flowers are singly in leaf axils and at branch ends. Petals are four, pale to bright yellow, 0.6–2 cm long, 0.4–1.7 cm wide, and broadly obovate. Shoot of this species is edible (Kew Royal Botanic Garden).

Nutritional Facts

Proximate Composition of Shoot (% DM)

Fat	CHO[a]	Protein
2.0	6.8	9.76

[a] Carbohydrate.

Minerals (% or mg/100 g DM)

P (%)	N2 (%)	K (%)	Na (%)	Fe (%)	Mg (mg)	Cu (mg)	Zn (mg)
0.98	1.56	0.13	0.02	0.75	0.82	0.09	0.33

Source: Data from Jain et al. (2011).

Limnophila aromaticoides (Yuen P. Yang & S.H. Yen)

Order: Scrophulariales

Family: Scrophulariaceae

Common name: Rice paddy plant, rice paddy herb.

Distribution: Native to Southeast Asia, North America.

Habitat: Swamps, water, or marshy places.

Description: It is a smooth herb with stems that are stout, erect, simple, 30–60 cm in length, and rarely branched above. Leaves are linear-oblong or oblong-lanceolate, 2–6 cm in length, 0.5–1 cm in width, opposite and whorled, with pointed tip, rounded and clasping base, and toothed margins. Flowers are pink or pale purple, and borne singly or in whorls in inflorescences at the axils of the leaves. Calyx is about 4 mm long and corolla is 1.2 cm long. The leafy stems of this species are edible (Yang and Yen, 1997).

Nutritional Facts

Proximate Composition (g/100 g DM)

Moisture[a]	CHO[b]	Protein	Fat	Fiber	Ash	Energy[c]
94.9	44.6	10.0	3.3	16.7	13.3	248

[a] g/100 g WW.
[b] Carbohydrate.
[c] kcal/100 g DW.

Minerals (per g or 100 g DM)

Na (mg/g)	K (mg/g)	Mg (mg/g)	Ca (mg/100 g)	Fe (mg/100 g)	Zn (µg/g)	Cu (µg/g)
10.5	21.8	3.13	11.9	11.3	216.4	61.3

Vitamins (µg/g WW)

Asc	γT	αT	Thi	Rib
844.6	7.0	14.8	16.6	13.7

Asc, ascorbic acid; γT, γ-tocopherol; αT, α-tocopherol; Thi, thiamin; Rib, riboflavin.

Source: Data from Ng et al. (2012).

Ipomoea aquatica (Forssk.)

Order: Solanales

Family: Convolvulaceae

Common name: Water spinach, river spinach.

Distribution: Throughout the tropical and subtropical regions of the world.

Habitat: Semiaquatic; water or on moist soil.

Description: Its stems are 2–3 m long, rooting at the nodes, and they are hollow and can float. Leaves are sagittate (arrowhead shaped) to lanceolate, 5–15 cm long, and 2–8 cm broad. Flowers are showy and funnel form like morning glory blooms 3–5 cm in diameter. They are solitary or in few-flowered clusters at leaf axils. Petals are usually white in color with a mauve center. The tender shoot and leaves of this species are edible.

Nutritional Facts

Proximate Composition of Leaves (% WW)

Moisture[a]	Ash	Protein	Lipid	Fiber	CHO[b]	Energy[c]
72.83	10.83	6.30	11.00	17.67	54.20	300.94

[a] % DW.

[b] Carbohydrate.

[c] kcal/100 g.

Minerals (mg/100 g DM)

K	Na	Ca	Mg	P	Cu	Fe	Mn	Zn	Co
5458	135	417	302	109	0.36	210	2.14	2.47	0.02

Source: Data from Umar et al. (2007).

Class: Polypodiopsida (Leptosporangiate Ferns)

Marsilea quadrifolia (Linnaeus)

Phylum: Magnoliophyta

Class: Polypodiopsida

Order: Salvinales

Family: Marsileaceae

Common name: Four-leaf clover, European water clover.

Distribution: Central and southern Europe, Caucasia, western Siberia, Afghanistan, southwest India, China, Japan, and North America.

Habitat: Shallow water of lakes and ponds, quiet sections of rivers and streams, wet shores.

Description: This is an aquatic fern bearing a four-parted leaf. Leaves are floating in deep water or erect in shallow water or on land. Leaflets obdeltoid, to 18 mm long, glaucous; petioles to 20 cm long. The leaves and seed of this species are edible.

Nutritional Facts

Proximate Composition (DM)

Protein (%)	Ash (%)	Beta-carotene (mg/100 g)
36.2	11.0	2.6

Minerals (% DM)

Na	K	Ca	P
1.4	1.8	0.5	0.5

Source: Data from Devanji et al. (1993).

66

Marsilea minuta (L.)

Order: Salvinales

Family: Marsileaceae

Common name: Dwarf water clover, small water clover, Gelid waterklawer.

Distribution: Widely distributed in tropical Africa and Asia (Mani, 2013).

Habitat: Shallow pools, at the edges of rivers, canals, and ditches and in rice fields; temporarily flooded places.

Description: It is a floating form. Stipe is up to 26 cm long, slender, hairless. Leaflets are up to 15–28 × 10–29 mm, broadly obovate, outer margin rounded, entire to shallowly irregular and hairless.

The shoot of this species is edible (Flora of Zimbabwe).

Nutritional Facts

Proximate Composition of Shoot (% DM)

Fat	CHO[a]	Protein
4.0	4.8	8.0

[a] Carbohydrate.

Minerals (% or mg/100 g DM)

P (%)	N2 (%)	K (%)	Na (%)	Fe (mg)	Mg (mg)	Cu (mg)	Zn (mg)
0.94	1.28	0.37	0.02	0.68	0.71	0.09	0.21

Source: Data from Jain et al. (2011).

67

Ceratopetris thalictroides (L.) (Brogn.)

Order: Polypodiales

Family: Pteridaceae

Common name: Water sprite, Indian fern, water fern, oriental water fern, water horn fern

Distribution: Cosmopolitan; pantropical; Asia, Australia, and America.

Habitat: Still or slow-moving waters; paddy fields, ponds.

Description: Plants of this species are usually rooted in mud, very variable in size and appearance. Scales on rhizome are peltate, thin, translucent, and pale brown. Stipes measure 3–15 mm diameter in mature plants. The fronds of this plant are used as vegetables.

Nutritional Facts

Proximate Composition (g/100 g DM)

Moisture[a]	CHO[b]	Protein	Fat	Fiber	Ash	Energy[c]
92.6	37.0	21.2	2.1	17.7	13.0	252

[a] g/100 g WW.
[b] Carbohydrate.
[c] kcal/100 g DM.

Minerals (per g or 100 g DM)

Na (mg/g)	K (mg/g)	Mg (mg/g)	Ca (mg/100 g)	Fe (mg/100 g)	Zn (µg/g)	Cu (µg/g)
3.3	24.0	2.89	11.0	7.1	70.3	113.0

Vitamins (µg/g WW)

Asc	γT	αT	Thi	Rib
551.8	12.2	3.5	9.0	8.9

Asc, ascorbic acid; γT, γ-tocopherol; αT, α-tocopherol; Thi, thiamin; Rib, riboflavin.
Source: Data from Ng et al. (2012).

Class: Liliopsida (Monocotyledons)

Dioscorea rotundata (Poir.)

Phylum: Magnoliophyta

Class: Lilliopsida

Order: Dioscoreales

Family: Discoreaceae

Common name: White yam.

Distribution: West Africa, including countries such as Ivory Coast, Ghana, and Nigeria.

Habitat: Moist soils, wetlands.

Description: A vigorous vine with a cylindrical stem, twining to the right, usually spiny and with dark glossy green leaves. Its foliage may have a purplish waxy bloom. Spines are very common, especially on the lower and larger stems. Stems may be somewhat striated vertically. The length of the vine varies, but heights of 10–12 m are easily accomplished. Length of leaves varies from 4 to 20 cm. Shapes of leaves are ovate, cordate, or almost orbicular. This species usually produces male and female flowers on separate plants. The male flowers are 1–3 mm in diameter, sessile, and borne on spikes subtended by small bracts. Female flowers occur on axillary spikes and are about 0.5 cm long. The tubers of this species are edible (Ecocrop FAO).

69

Nutritional Facts

Proximate Composition (% DM)

Moisture	54.50
Ash	1.40
Crude fat	2.70
Crude protein	0.087
Crude fiber	0.70
CHO[a]	40.61
Energy[b]	731.75

[a] Carbohydrate.
[b] kJ/100 g sample.

Minerals (mg/100 g)

Na	185.15
P	209.13
Ca	132.02
Mg	45.90
Fe	81.85
Cu	10.06
Zn	5.46
P	54.00

Source: Data from Alinnor and Akalezi (2010).

Commelina benghalensis (L.)

Order: Commelinales

Family: Commelinaceae

Common name: Benghal dayflower, tropical spiderwort.

Distribution: Perennial herb native to tropical Asia and Africa, Hawaii, the West Indies, and both coasts of North America; Alabama, California, Georgia.

Habitat: Wet locations.

Description: The stem of this annual, perennial species is ascending, can extend more than 1 m, and is capable of rooting from nodes. The oval leaf blades are 3–7 cm long by 1–4 cm wide. Leaves often have reddish hairs toward the tip. This plant produces both aerial and underground flowers. Aboveground flowers are lilac to blue and very small (3–5 mm per petal). Belowground flowers are white and very small. The roots and tubers of this species are used as a food source (FAO).

Nutritional Facts

Proximate Composition of Shoot (% DM)

Fat	CHO[a]	Protein
1.0	5.0	9.4

[a] Carbohydrate.

Minerals (% DM)

P	N2	K	Na
0.82	1.50	0.34	0.02

Source: Data from Jain et al. (2011).

Monochoria vaginalis (Burm. f.) (C. Presl ex Kunth)

Order: Commelinales

Family: Pontederiaceae

Common name: Heartleaf false pickerelweed, oval leaf pondweed.

Distribution: Native to much of Asia and across many of the Pacific islands.

Habitat: Freshwater pools, stagnant backwaters, mudflats, swampy places, ditches.

Description: This is an annual or perennial herb 10–50 cm tall and stemless.

The shiny green leaves are up to about 12 cm long and 10 cm wide. They are linear or lanceolate, and in older plants, they are ovate-oblong to broadly ovate, sharply acuminate with a heart-shaped or rounded base, shiny, deep green. The inflorescence bears 3–25 flowers, which open underwater and all around the same time. Each has six purple-blue sepals just over a centimeter long. The leaves and flowers of this species are edible.

Nutritional Facts

Proximate Composition (g/100 g DM)

Moisture[a]	CHO[b]	Protein	Fat	Fiber	Ash	Energy[c]
97.5	38.5	19.6	0.9	11.3	13.2	240

[a] g/100 g WW.

[b] Carbohydrate.

[c] kcal/100 g DM.

Minerals (mg/g)

Na (mg/g)	K (mg/g)	Mg (mg/g)	Ca (mg/100 g)	Fe (mg/100 g)	Zn (μg/g)	Cu (μg/g)
18.8	20.2	1.64	4.4	33.1	37.8	67.1

Vitamins (μg/g WW)

Asc	γT	αT	Thi	Rib
438.1	ND	1.1	ND	2.3

Asc, ascorbic acid; γT, γ-tocopherol; αT, α-tocopherol; Thi, thiamin; Rib, riboflavin; ND, no data.

Source: Data from Ng et al. (2012).

Lemna minor (Linnaeus)

Order: Alismatales

Family: Araceae

Common name: Common duckweed, lesser duckweed.

Distribution: Subcosmopolitan distribution, native, throughout most of Africa, Asia, Europe, and North America.

Habitat: Freshwater ponds and slow-moving streams.

Description: It is a floating plant, with one, two, or three leaves, each with a single root hanging in the water; as more leaves grow, the plants divide and become separate individuals. The root is 1–2 cm long. The leaves are oval, 1–8 mm long and 0.6–5 mm broad, light green, with three (rarely five) veins,

and small air spaces to assist flotation. It propagates mainly by division, and flowers are rarely produced; when produced, they are about 1 mm diameter, with a cup-shaped membranous scale. The leaves of this species are edible.

Nutritional Facts

Proximate Composition (% DM)

Protein	Ash	Fiber
17.86	1.61	11.82

Minerals (% or mg/100 g DM)

Ca (%)	P (%)	K (%)	Na (%)	Mg (%)	Fe (mg)	Mn (mg)	Mo (mg)
2.58	0.17	1.20	0.755	0.457	248	103	13.2

Source: Data from Linn et al. (1975).

Orontium aquaticum (Linnaeus)

Order: Alismatales

Family: Araceae

Common name: Golden club.

Distribution: Native range: Eastern United States.

Habitat: Ponds, streams, and shallow lakes.

Description: It is a rhizomatous marginal aquatic perennial. It grows 1–2 ft tall and spreads indefinitely by stout, slowly creeping rhizomes. Long-stalked elliptic dark bluish green leaves (to 30 cm long) are submerged, floating, or aerial. Inflorescence is most notable for having an extremely small, almost indistinguishable sheath surrounding the spadix, which is 20 cm tall.

The root and seed of this species are edible.

Nutritional Facts

Proximate Composition (% DM)

Ash	Fat	Protein
7.38	14.88	49.6

Amino Acids (% DM)

Arginine	3.17
Histidine	1.02
Isoleucine	2.32
Leucine	4.30
Methionine	0.84
Phenylalanine	2.80
Threonine	2.26
Valine	2.55

Source: Data from Boyd (1968b).

Colocasia esculenta (L.) (Schott)

Order: Alismatales

Family: Araceae

Common name: Taro, elephant ears.

Distribution: Native to the lowland wetlands of Malaysia; western Australia.

Habitat: Swampy areas.

Description: It is an emergent aquatic and semiaquatic plant. Rhizomes are of different shapes and sizes. Leaves up to 40 × 24.8 cm, sprouts from rhizome, dark green above and light green beneath, triangular-ovate, subrounded, and mucronate at apex; tip of the basal lobes rounded or subrounded. Spadix about 3/5 as long as the spathe, flowering parts up to 8 mm in diameter. Female portion at the fertile ovaries intermixed with sterile white ones. The corms of this species are edible.

Nutritional Facts

Proximate Composition of Leaves (% DM)

Moisture	Ash	Fiber	Protein	Fat	CHO[a]	Energy[b]
8.00	10.00	1.00	5.57	4.5	78.93	378.5

[a] Carbohydrate.
[b] kcal/100 g.

Minerals (mg/100 g)

K	Na	Ca	Fe
94	9.8	2.15	4.0

Source: Data from Yirankinyuki and Lamayi (2013).

Sagittaria latifolia (Willd.)

Order: Alismatales

Family: Alismataceae

Common name: Duck potato, wapato, arrowhead.

Distribution: Native to southern Canada, United States, as well as Mexico, Central America, Colombia, Venezuela, Ecuador, and Cuba; Puerto Rico, Bhutan, Australia, and much of Europe (France, Spain, Italy, Romania, Germany, Switzerland, the Czech Republic, and European Russia).

Habitat: Swampy areas; emergent marshes, ponds, lakes, sloughs, lagoons, and wet areas.

Description: It is a vigorous, deciduous, marginal aquatic perennial that typically grows 10 m tall. Each plant produces a rosette of leaves and an inflorescence on a long rigid scape. Immersed leaves (to 30 cm long) are typically broadly sagittate (arrowhead shaped). Submerged leaves are often much narrower (linear to ovate). Inflorescence is a raceme composed of large flowers whorled by threes.

Usually divided into female flowers on the lower part and male on the upper, although dioecious individuals are also found. Three round, white petals and three very short curved, dark green sepals. Young unfurling leaves and stalk and white or bluish tubers of this species are edible.

Nutritional Facts

Proximate Composition (% DM)

Ash	Fat	Protein
4.23	16.62	42.8

Amino Acids (% DM)

Arginine	2.14
Histidine	1.14
Isoleucine	1.46
Leucine	1.87
Methionine	0.56
Phenylalanine	T
Threonine	1.57
Valine	1.80

T, trace.
Source: Data from Boyd (1968b).

Sagittaria cuneata (E. Sheld.)

Order: Alismatales

Family: Alismataceae

Common name: Arumleaf arrowhead, duck potato.

Distribution: Native to much of North America, including most of Canada as well as the western and northeastern United States (New England, Great Lakes, Great Plains, Rocky Mountains, Great Basin, and Pacific Coast states).

Habitat: Slow-moving and stagnant water bodies such as ponds, lakes, and small streams.

Description: It is a perennial herb growing from a white or blue-tinged tuber. Leaves are variable in shape, many of them sagittate (arrow shaped) with two smaller, pointed lobes opposite the tip. Leaf blades are borne on very long petioles. The plant is monoecious, with individuals bearing both male and female flowers.

The inflorescence that rises above the surface of the water is a raceme made up of several whorls of flowers, the lowest node bearing female flowers and upper nodes bearing male flowers. The flower is up to 2.5 cm wide with white petals. Large starchy tubers of this species are edible after baking or boiling.

Nutritional Facts

Proximate Composition (% DM)

Protein	Ash	Fiber
21.81	1.91	17.34

Minerals (% or mg/100 g DM)

Ca (%)	P (%)	K (%)	Na (%)	Mg (%)	Fe (mg)	Mn (mg)	Mo (mg)
0.78	0.55	2.82	0.391	0.354	1904	161	28.2

Source: Data from Linn et al. (1975).

Sagittaria rigida (Pursh)

Order: Alismatales

Family: Alismataceae.

Common name: Sessile-fruited arrowhead, stiff arrowhead.

Distribution: Native to Canada and the United States, UK.

Habitat: Shallow water, swamps, marshes, streams, lake borders, and ponds.

Description: It is an erect or floating perennial, emergent aquatic, 10–60 cm tall. Flower is white, three-parted, 1.5–2.5 cm wide. Inflorescence is of two to eight whorls of flowers with 1 unbranched stalk. Upper male flowers large and stalked, and lower female flowers are smaller and stalkless. Leaf is usually broadly lance shaped, sometimes with small lobes at the base. Tubers, laterals, young leaves, and shoots of the flower stalks of this species are edible (Plants for a Future, 1996–2012c).

Nutritional Facts

Proximate Composition (% DM)

Protein	Ash	Fiber
14.78	2.27	23.69

Minerals (% or mg/100 g DM)

Ca (%)	P (%)	K (%)	Na (%)	Mg (%)	Fe (mg)	Mn (mg)	Mo (mg)
0.99	0.31	1.82	0.243	0.432	2083	236	30.2

Source: Data from Linn et al. (1975).

77

Vallisneria americana (Michx.)

Order: Alismatales

Family: Hydrocharitaceae

Common name: Wild celery, water celery, tape grass, eelgrass.

Distribution: Americas, Iraq, China, Japan, Korea, India, Papua New Guinea, Philippines, Australia, Canada, United States, Mexico, Guatemala, Honduras, Cuba, the Dominican Republic, Haiti, and Venezuela.

Habitat: Lakes and streams; grows underwater.

Description: The plants of this species are long, limp, flat, and have a green midridge. They spread by runners and sometimes form tall underwater meadows. Eelgrass leaves arise in clusters from their roots. They are 2.5 cm wide and can be several meters long. The leaves, which are up to 1 m long, have rounded tips and definite raised veins. Single white female flowers grow to the water surface on very long stalks. Leaves of this species are edible (Center for Aquatic and Invasive Plants).

Nutritional Facts

Proximate Composition (% DM)

Protein	Ash	Fiber
15.15	3.10	27.32

Minerals (% or mg/100 g DM)

Ca (%)	P (%)	K (%)	Na (%)	Mg (%)	Fe (mg)	Mn (mg)	Mo (mg)
1.82	0.16	3.75	0.571	0.298	323	176	12

Source: Data from Linn et al. (1975).

Potamogeton amplifolius (Tuckerman)

Order: Alismatales

Family: Potamogetonaceae

Common name: Large-leaf pondweed, broad-leaved pondweed.

Distribution: Through much of North America.

Habitat: Lakes, ponds, and rivers, often in deep water.

Description: This perennial plant grows from rhizomes and produces a very slender, cylindrical, sometimes spotted stem up to a meter or so long. Leaves take two forms. Submersed leaves are up to 20 cm long by 7 cm wide and may be folded along their midribs. Floating leaves are up to 10 cm long by 5 cm wide, leathery in texture, and borne on long petioles. The inflorescence is a spike of many flowers rising above the water surface on a thick peduncle. The tuber of this species is edible (Naturalist.org).

Nutritional Facts

Proximate Composition (% DM)

Protein	Ash	Fiber
14.36	2.39	15.72

Minerals (% or mg/100 g DM)

Ca (%)	P (%)	K (%)	Na (%)	Mg (%)	Fe (mg)	Mn (mg)	Mo (mg)
1.01	0.83	2.91	0.172	0.232	1651	505	41.0

Source: Data from Linn et al. (1975).

Potamogeton pectinatus (Linnaeus) (= *Stuckenia pectinata*)

Order: Alismatales

Family: Potamogetonaceae

Common name: Fennel-leaved pondweed, sago pondweed.

Distribution: Cosmopolitan on all continents except Antarctica.

Habitat: Ponds, rivers, canals, ditches, etc.; fresh and brackish water.

Description: This species has long narrow linear leaves that are less than 2 mm wide; each is composed of two slender parallel tubes. The flowers are wind pollinated and the seeds float. Tubers that are rich in starch are formed on the rhizomes. Leaves and stems of this species are edible.

79

Nutritional Facts

Proximate Composition (% DM)

Protein	Ash	Fiber
14.05	3.22	15.64

Minerals (% or mg/100 g DM)

Ca (%)	P (%)	K (%)	Na (%)	Mg (%)	Fe (mg)	Mn (mg)	Mo (mg)
3.76	0.34	1.99	0.107	0.200	1171	535	21.0

Source: Data from Linn et al. (1975).

Potamogeton richardsonii (A. Bennett) (Rydberg)

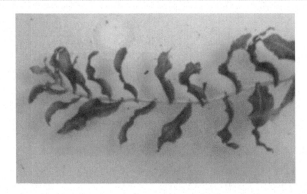

Order: Alismatales

Family: Potamogetonaceae

Common name: Richardson's pond-weed, redhead pondweed.

Distribution: Native to much of northern North America, including all of Canada and the northern and western United States.

Habitat: Lacustrine (in lakes or ponds), riverine (in rivers or streams).

Description: This perennial herb grows a narrow, mostly unbranched stem from a mat of rhizomes in the substrate. It reaches about a meter in maximum length. Leaves of this species are all submerged underwater. They are alternately arranged. Leaves are up to 13 cm long and about 3 cm wide and are olive green, ovate-lanceolate to narrowly lanceolate, not arcuate, base rounded, margins entire to crispate, and apex not hood-like. Inflorescences are emersed, unbranched. The small red berry of this species is edible.

Nutritional Facts

Proximate Composition (% DM)

Protein	Ash	Fiber
11.20	2.62	19.11

Minerals (% DM)

Ca (%)	P (%)	K (%)	Na (%)	Mg (%)	Fe (mg)	Mn (mg)	Mo (mg)
4.03	0.17	1.33	0.139	0.500	218	109	11.6

Source: Data from Linn et al. (1975).

Typha latifolia (Linnaeus)

Order: Poales (Cyperales)

Family: Typhaceae

Common name: Bulrush, common bulrush, broadleaf cattail, common cattail.

Distribution: Native to North and South America, Europe, Eurasia, and Africa; Canada, Yukon, and Northwest Territories, and in the United States.

Habitat: Flooded areas, freshwater, slightly brackish marshes.

Description: It is a marginal aquatic perennial growing to 3 m high. Narrow, upright, sword-like, linear, mostly basal, green leaves (210 cm long and 4 cm broad) are present. A stiff, unbranched central flower stalk that typically rises equal to or slightly less than the height of the leaves. Plants are monoecious, with each flower stalk being topped by two sets of minute flowers densely packed into a cylindrical inflorescence. Yellowish male (staminate) flowers are located at the top of the inflorescence, and greenish female (pistillate) flowers are located underneath. The rhizomes are edible after cooking and removing the skin, while peeled stems and leaf bases can be eaten raw or cooked. Young flower spikes are edible as well (Missouri Botanical Garden).

Nutritional Facts

Proximate Composition (% DM)

Ash	Protein	Fat	Cellulose	Energy[a]
6.9	10.3	3.9	33.2	3.7

[a] kcal/g.

Source: Data from Boyd (1968b).

81

Zizania aquatica (Linnaeus)

Order: Poales

Family: Poaceae

Common name: Wild rice, Canada rice, Indian rice, water oats.

Distribution: North America, China (Memidex).

Habitat: Shallow water in small lakes and slow-flowing streams.

Description: It is a very large, ribbon-like grass with stems 270 cm tall, thick and spongy. Leaf blades are strap-like, 90–120 cm long to 5 cm wide and smooth with margins sharply toothed. Inflorescence is erect, very large, at stem tip, to 260 cm long and 30 cm across, spreading branches and branchlets, lower branchlets drooping (male flowers), upper branchlets pointing stiffly upward (female flowers). Spikelets and flowers are numerous. The seeds are edible and well known (Center for Aquatic and Invasive Plants, University of Florida).

Nutritional Facts

Proximate Composition (% DM)

Moisture	7.9–11.2
Protein	12.4–15.0
Fat	0.5–0.8
Ash	1.2–1.4

Crude fiber	0.6–1.1
Total carbohydrate	72.3–75.3

Amino Acids (% of Protein)

Alanine	6.2
Arginine	8.2
Aspartic acid	10.5
Cystine	0.3
Giutamic acid	19.3
Glvcine	4.9
Histidine	2.8
Isoleucine	4.4
Leucine	7.4
Lvsine	4.6
Methionine	3.0
Phenylalanine	5.1
Proline	3.9
Serine	5.4
Threonine	3.2
Tyrosine	4.0
Valine	7.0

Source: Data from Anderson (1976).

Fatty Acids (g/100 g)

Total Fat	SFA	PUFA	MUFA
1.1	0.2	0.7	0.2

Source: Data from U.S. Department of Agriculture, National Nutrient Database for Standard Reference, Standard Release 26.

Zizania latifolia (Griseb.) (Turcz. ex Stapf.)

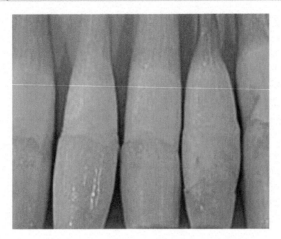

Order: Poales

Family: Poaceae

Common name: Manchurian wild rice.

Distribution: Europe: Eastern. Asia—temperate: Siberia, Soviet Far East, China, and eastern Asia. Asia—tropical: India, Indo-China, and Malaysia. Australasia: New Zealand. Pacific: North-central.

Habitat: Shallow water of lake margins and swamps.

Description: It is a tall, rhizomatous, perennial grass that grows up to 3 m tall. Leaf blades are 30–100 cm long and 20–30 mm wide; coriaceous. Leaf blade surface is scaberulous; rough on both sides. Leaf blade margins are cartilaginous and apex is attenuated. Inflorescence is monoecious, with male and female spikelets in the same inflorescence. Inflorescence is a panicle. The flowers, root, stem, and seed of this species are edible (http://www.rnzih.org.nz/pages/nppa_057.pdf).

Nutritional Facts

Proximate Composition of Shoot (% DM)

Fat	CHO[a]	Protein
1.0	13.8	8.13

[a] Carbohydrate.

Minerals (% or mg/100 g DM)

P (%)	N2 (%)	K (%)	Na (%)	Fe (mg)	Mg (mg)	Cu (mg)	Zn (mg)
0.78	1.30	0.46	0.02	0.85	3.34	0.12	4.71

Source: Data from Jain et al. (2011).

Oryza sativa (Linnaeus)

PLANT

SEED

Order: Poales

Family: Poaceae

Common name: Asian rice.

Distribution: Temperate East Asia, upland areas of Southeast Asia, and high elevations in South Asia.

Habitat: Aerobic moist soil.

Description: It is generally an annual grass, although some varieties are perennial. Plants typically grow in a tuft (clump) of upright culms (stems) up to 2 m or more tall, with long, flat leaf blades. Flowers grow on broad, open terminal panicles (branched clusters). Oblong spikelets, which each contain a single flower (that develops into a single kernel of grain), are sparse along the

84

stem rather than forming dense clusters. Harvested edible kernel, known as a rice paddy, is enveloped in a hull or husk that is removed during milling (EOL).

Nutritional Facts

Proximate Composition (% DM)

Moisture	Protein	Lipid	Fiber	CHO[a]	Ash
9.20	10.68	2.04	0.84	75.69	1.55

[a] Carbohydrate.

Minerals (mg/100 g)

Na	K	Ca	Mg	P
140	315	122	118	354

Fatty Acids (% of total fatty acids)

Saturated	23.55
Unsaturated	76.45
Unsaturated/saturated ratio	3.25

Source: Data from Al-Bahranay (2002).

Oryza rufipogon (Griff.)

Order: Poales

Family: Poaceae

Common name: Brownbeard rice, wild rice, red rice.

Distribution: Widely distributed tropical plant. Asia (Afghanistan, Bangladesh, Cambodia, China, India, Indonesia, Iran, Iraq, Democratic People's Republic of Korea, Laos, Malaysia, Peninsular Malaysia, Myanmar, Nepal, Pakistan, Philippines, Sri Lanka, Thailand, Vietnam), Africa (Egypt, Senegal, Swaziland, Tanzania), North America (United States), Central America, South America (Brazil, Colombia, Ecuador, Guyana, Peru, Venezuela, Oceania), Australia (Australia, Queensland, Papua New Guinea) (Watve, 2013).

Habitat: Shallow water, irrigated fields, pools, ditches, and sites with stagnant or slow running water.

Description: It is a perennial species with rhizomes that are elongated. Leaf sheaths are smooth and glabrous on surface. Leaf sheath auricles are erect. Leaf blade margins are scabrous. Inflorescence is a panicle. Spikelets are appressed and solitary. Fertile spikelets are pedicelled. Basal sterile florets are similar, barren, without significant palea. This species is recognized as an

85

aggressive weed of rice. It is classified as a weed because the seeds can go dormant, the grain coat is red, and the spikelet shatters. It yields edible rice (Fasahat et al., 2012; Biodiversity India).

Nutritional Facts

Proximate Composition (DM)

Moisture content (%)	12.15
Protein (g per 100 g) ($N \times 5.95$)	8.0
Fat content (%)	2.2
Amylose (%)	25.2

Source: Data from Fasahat et al. (2012).

Eleocharis dulcis (Burm. f.) (Trin. ex Hensch.)

Order: Poales

Family: Cyperaceae

Common name: Chinese water chestnut.

Distribution: Native to Asia (China, Japan, India, Philippines, etc.), Australia, tropical Africa, and various islands of the Pacific and Indian Oceans.

Habitat: Marshy land and shallow waters.

Description: This species is leafless and perennial. The plants grow from 50 to 200 cm tall and are deep shining green in color. They produce many flowers that are very small and occur on the tips of the culms. The plants have elongated stolons with a tuber attached to them at their bottom. The plants produce two types of tubers: the first type for propagation and the second for storage. The second type of corms (tubers) is the edible water chestnut (Thulaja, 2005).

Nutritional Facts

Proximate Composition (%) and Minerals (mg/100 g WW)

Ash	Sugar	Starch	Ca	K	P
0.84	10.28	6.12	8.70	409	49

Source: Data from Wong (1964).

Moisture and Protein (% WW)

Moisture	Protein
81.1	13.9

Source: Data from Hasan and Chakrabarti (2009).

Cyperus esculentus (Linnaeus)

Order: Poales

Family: Cyperaceae

Common name: Chufa sedge, nut grass, yellow nut sedge, tiger nut sedge, earth almond.

Distribution: Native to most of the western hemisphere as well as southern Europe, Africa, Madagascar, the Middle East, and the Indian subcontinent.

Habitat: Moist soils, wetlands.

Description: It is an annual or perennial plant, growing to 90 cm tall, with solitary stems growing from a tuber. The stems are triangular in section and bear slender leaves 3–10 mm wide. The spikelets of the plant are distinctive, with a cluster of flat, oval seeds surrounded by four hanging, leaf-like bracts positioned 90° from each other. They are 5–30 mm long and linear to narrowly elliptic with pointed tips and 8–35 florets. The color varies between straw colored to gold-brown. The plant foliage is very tough and fibrous. Tubers of this species are edible.

Nutritional Facts

Proximate Composition (% DM)

Moisture	Ash	Protein	Fats	CHO[a]	Fiber	Energy[b]
9.74	1.79	3.94	27.54	41.59	15.60	429.18

[a] Carbohydrate.

[b] kcal/100 g.

Source: Data from Monago and Uwakwe (2009).

Amino Acids (mg/100 g)

	Wet Sample	Dry Sample
Lysine	4.16	4.78
Proline	2.02	2.34
Histidine	1.94	2.32
Glycine	4.13	3.19
Arginine	18.29	20.17
Alanine	3.17	3.55
Aspartate	5.73	6.86
Cysteine	2.05	2.32
Threonine	2.77	3.00
Valine	3.14	2.50
Serine	2.27	2.37
Methionine	0.73	0.89
Glutamic acid	7.25	7.98
Leucine	3.71	4.01
Isoleucine	2.07	1.79
Tyrosine	0.81	0.96
Phenylalanine	2.37	2.37

Source: Data from Nwaoguikpe and Nwazue (2010).

87

Carex stricta (Lam.)

Order: Poales

Family: Cyperaceae

Common name: Upright sedge, tussock sedge.

Distribution: Native to central United States; eastern North America (Missouri Botanical Garden).

Habitat: Moist marshes, forests, and alongside bodies of water.

Description: It is a rhizomatous evergreen sedge that grows in dense tussocks (clumps) to 90 cm tall and 60 cm wide. It is an emergent aquatic. Narrow, glaucous, grass-like leaves (to 1 cm wide) grow in dense clumps. Older leaves turn straw brown as they die, and build up around the base of each clump surrounding the newer yellowish green leaves. Leaves of this species are eaten raw (Missouri Botanical Garden).

Nutritional Facts

Proximate Composition (% DM)

Protein	Ash	Fiber
9.94	0.90	29.41

Minerals (% or mg/100 g DM)

Ca (%)	P (%)	K (%)	Na (%)	Mg (%)	Fe (mg)	Mn (mg)	Mo (mg)
0.28	0.17	0.52	0.020	0.140	861	185	15.2

Source: Data from Linn et al. (1975).

Alpingia galanga (L.) (Willd.)

Order: Zingiberales

Family: Zingiberaceae

Common name: Blue ginger, Thai ginger.

Distribution: Native to South Asia and Indonesia; Malaysia, Laos, and Thailand.

Habitat: Wetlands.

Description: It is a perennial herb with adventitious roots. Rhizomes are cylindrical, branched, and 2–8 cm in diameter. These rhizomes are longitudinally ridged with prominent warts marked with fine annulations. Scaly leaves arranged circularly, extremely reddish brown, and internally orange-yellow. Rhizome of this species has edible and medicinal values (Chudiwal et al., 2010).

Nutritional Facts

Proximate Composition of Shoot (% DM)

Fat	CHO[a]	Protein
1.0	4.4	2.6

[a] Carbohydrate.

Minerals (% or mg/100 g DM)

P (%)	N2 (%)	K (%)	Na (%)	Fe (mg)	Mg (mg)	Cu (mg)	Zn (mg)
0.58	0.40	0.33	0.02	1.25	3.10	0.03	0.45

Source: Data from Jain et al. (2011).

Hedycium coronarium (J. Koenig)

Order: Zingiberales

Family: Zingiberaceae

Common name: White ginger lily, ginger lily, white butterfly ginger lily, garland flower.

Distribution: Native to India; temperate regions of North America and Europe; Caribbean and tropical and subtropical areas worldwide (FLORIDATA).

Habitat: Wetlands; moist places along streams and on forest edges; rich soil with adequate moisture.

Description: It is a tropical perennial. Its green stalks grow from thick rhizomes to a height of 0.9–2.1 m. Leaves are lance shaped and sharp pointed, 20–61 cm long and 5–12.7 cm wide, and arranged in two neat ranks that run the length of the stem. The stalks are topped with 15.2–30.5 cm long clusters of wonderfully fragrant white flowers that look like butterflies. Rhizomes of this species have edible and medicinal values (FLORIDATA).

Nutritional Facts

Proximate Composition of Shoot (% DM)

Fat	CHO[a]	Protein
2.0	10.0	4.63

[a] Carbohydrate.

Minerals (% DM)

P	N2	K	Na
0.92	0.92	0.06	0.02

Source: Data from Jain et al. (2011).

Etlingera elatoir (Jack) (R.M. Sm.)

Order: Zingiberales

Family: Zingiberaceae

Common name: Torch ginger, ginger flower, red ginger lily, torch lily, wild ginger.

Distribution: Native to parts of Southeast Asia; India to the Pacific islands; pantropical distribution (National Tropical Botanical Garden).

Habitat: Near rivers, wet habitats.

Description: It is a rhizomatous perennial herb up to 6 m tall. Leaf blades are green, hairless, lanceolate in shape, and up to 81 cm long. Pseudostems (formed by the leaf sheaths) emerge from underground rhizomes and are tall and arching. Large, torch-like, up to 1.5 m tall flower stalks emerge from fleshy underground rhizomes. Inflorescences have waxy, red to pink, white-edged bracts and are pinecone shaped with a skirt of larger bracts. Individual flowers emerge from between the colorful bracts and have a dark red labellum (lip petal) with a bright yellow margin (Hawaiian Plants and Tropical Flowers; Elden Project).

Nutritional Facts

Proximate Composition (g/100 g DM)

Moisture[a]	CHO[b]	Protein	Fat	Fiber	Ash	Energy[c]
94.9	36.4	16.2	1.6	19.8	17.6	225

[a] g/100 g WW.

[b] Carbohydrate.

[c] kcal/100 g DW.

91

Minerals (mg/g)

Na (mg/g)	K (mg/g)	Mg (mg/g)	Ca (mg/100 g)	Fe (mg/100 g)	Zn (µg/g)	Cu (µg/g)
5.2	39.0	2.91	11.5	3.6	183.8	107.7

Vitamins (µg/g WW)

Asc	γT	αT	Thi	Rib
272.2	11.5	12.3	3.2	4.4

Asc, ascorbic acid; γT, γ-tocopherol; αT, α-tocopherol; Thi, thiamin; Rib, riboflavin.

Source: Data from Ng et al. (2012).

3
CRUSTACEANS

PRAWNS

Macrobrachium scabriculum Heller 1862

Phylum: Arthroipoda

Class: Crustacea

Order: Decapoda

Family: Palaemonidae

Common name: Goda river prawn.

Distribution: Eastern Africa, Madagascar, India, Bangladesh, Sri Lanka, and Sumatra (De Grave et al., 2013b).

Habitat: Fresh and brackish water of larger rivers and streams.

Description: This species has a massive rostrum of medium length and very developed claws, which are much more slender in males. The coloration background is more or less dark brown. A lighter dorsal line goes all the way from the rostrum to the telson. A half circle–shaped cream-colored line with black-brown marks the third segment of the abdomen. The legs and claws can show a slight blue shade. The external edges of the uropods are underlined with a cream-colored line. This species is omnivorous. The male measure 12 cm, the female 6–7 cm (Crusta-Fauna).

Nutritional Facts

Proximate Composition of Edible Portion (% DM)

Sex	Protein	CHO[a]	Lipid	Ash	Moisture[b]
Male	53.35	2.42	3.83	6.40	80.46
Female	54.24	2.04	4.28	6.05	79.06
Berried	46.74	1.69	3.76	6.42	81.02

[a] Carbohydrate.
[b] Wet weight.

Fatty Acids (% of Total Fatty Acids)

Saturated Fatty Acids (SFAs)

Fatty Acid	Female	Male
C12:0 lauric acid	0.83	0.91
C14:0 myristic	3.81	3.06
C15:0 pentadecyclic	1.55	1.18
C16:0 palmitic	27.64	17.34
C17:0 margaric	1.51	1.09
C18:0 stearic	9.67	6.42
C20:0 arachidic	0.82	0.40
Total	45.83	30.40

Monounsaturated Fatty Acids (MUFAs)

Fatty Acid	Female	Male
C16:1w5c ambrettolic	0.56	—
C17:1w8c	0.85	0.53
C18:1w9c oleic	21.04	14.46
C18:1w7c octadecenoic	4.81	3.87
Total	27.26	18.86

Polyunsaturated Fatty Acids (PUFAs)

Fatty Acid	Female	Male
C18:w6c linolenic	—	0.39
C20:4w6c arachidonic	4.75	4.84
Total	4.75	5.23

Source: Data from Dinakaran et al. (2010).

Macrobrachium idella idella Hilgendorf 1898 (= *Macrobrachium rude*)

Order: Decapoda

Family: Palaemonidae

Common name: River prawn.

Distribution: East Africa, Madagascar, Sri Lanka, India, and Bangladesh (De Grave et al., 2013c).

Habitat: Rivers and lakes.

Description: In this species, the second legs are covered by a very fine granulation that is often hidden beneath the pubescence. Both fingers have the cutting edge lined on either side by a row of tubercles that are often sharply pointed. The hepatic spine is very slightly lower than the antennal spine (Johnson, 1973).

Nutritional Facts

Proximate Composition (% DM)

	Protein	CHO[a]	Lipid
Male	52.44	1.96	3.76
Female	50.25	2.13	3.40

[a] Carbohydrate.

Source: Data from Athiyaman and Rajendran (2013).

Macrobrachium malcomsonii H. Milne-Edwards 1844

Order: Decapoda

Family: Palaemonidae

Common name: Monsoon river prawn.

Distribution: India, Bangladesh, Sri Lanka and Myanmar, Pakistan (De Grave, 2013b).

Habitat: Larger river systems.

Description: This species, which has a maximum size of 125 mm, is distinguished by its smaller size, relatively long and slender second legs, greater spinosity, and very short to moderately long rostrum. The rostral crest has 7–11 teeth. The distal, dorsal margin of the rostrum usually bears one or two teeth, but may be unarmed or bear as many as three or four teeth. The ventral margin of the rostrum has four to six teeth. The second legs of the adult male are fairly massive, the palm of which is slender and of uniform width throughout. Old males possess a very scabrous carapace. The posterior thoracic legs are somewhat slender (Johnson, 1973).

Macrobrachium rosenbergii De Man 1879

Order: Decapoda

Family: Palaemonidae

Common name: Giant freshwater prawn.

Distribution: Indo-Pacific region, northern Australia, and Southeast Asia (De Grave et al., 2013a).

Habitat: Tropical freshwater environments influenced by adjacent brackish water areas.

Description: Males of this species may reach a total length of 320 mm and females 250 mm. Body of the animal is usually greenish to brownish grey, sometimes more bluish, darker in larger specimens. Antennae are often blue, and chelipeds are blue or orange. Carapace is smooth and hard. Rostrum is long, slender, and somewhat sigmoid with 11–14 dorsal and 8–10 ventral teeth. Eyes are stalked. Thorax contains three pairs of maxillipeds, used as mouthparts, and five pairs of pereiopods (true legs). The first two pairs of pereiopods are chelate. Each pair of chelipeds are equal in size. The second chelipeds bear numerous spinules and are robust and slender (New, 2014).

Nutritional Facts

Proximate Composition, Amino Acids, and Fatty Acids (% DM)

	M. malcomsonii	*M. rosenbergii*
Protein	29.38	42.45
Carbohydrate	7.52	10.22
Lipid	3.38	4.49
Amino acids	21.18	30.79
Fatty acids	3.77	5.45
Moisture[a]	80.34	81.38
Ash	7.38	8.49

[a] Wet weight.

Amino Acids (g/100 g FW)

	M. malcomsonii	*M. rosenbergii*
Arginine	0.68	0.89
Histidine	1.79	2.38
Phenylalanine	0.54	0.77
Leucine	0.68	0.87
Tyrosine	1.68	2.23
Tryptophan	1.69	2.34
Methionine	2.03	2.37
Valine	0.68	0.95
Threonine	1.12	1.39
Glutamine	0.93	1.13
Glycine	1.13	1.79
Proline	0.88	1.12

Fatty Acids (mg/100 g FW)

	M. malcomsonii	*M. rosenbergii*
Palmitic acid (16:0)	641.18	835.35
Oleic acid (16:1)	589.92	875.18
Linoleic acid (18:2)	385.41	582.45
Linolenic acid (18:3)	130.14	142.17
Eicosapentaenoic acid (20:5)	385.14	540.10
Decosahexanoic acid (22:5)	103.45	138.00
Arachidic acid (20:0)	85.49	112.14
Arachidonic acid (20:4)	52.95	78.35

Source: Data from Rangappa et al. (2012).

Macrobrachium vollenhovenii Herklots 1857

Order: Decapoda

Family: Palaemonidae

Common name: African river prawn.

Distribution: Tropical West Africa from Senegal to Angola (Aquaristik-Elmshorn; http://www.wirbellose.de).

Habitat: Freshwater and low-salinity waters.

Description: It is a very large *Macrobrachium* species with characteristic flesh-colored scissors with blue fingers and an orange spot on the scissor hinge. The whole scissors leg is covered with small spines. Males possess significantly larger shears and an appendix masculina on the second pair of swimming legs, and females have a fülligerem abdomen. This species grows to a maximum length of 195 mm (maximum weight 83 g) and a carapace length of 38.0 mm (Aquaristik-Elmshorn).

Nutritional Facts

Proximate Composition of Edible Portion (% DM)

Protein	Fat	Ash	Fiber	Moisture	CHO[a]
71.37	7.33	10.31	0.45	10.71	2.42

[a] Carbohydrate.

Minerals (mg/100 g)

Ca	Mg	P	Zn
47.64	20.80	106.71	1.16

Amino Acids (mg/100 g)

Alanine	37.18
Arginine	39.55
Asparagine	68.35
Glutamine	108.03
Glycine	36.98
Histidine	13.46
Serine	14.56
Threonine	21.75
Leucine	62.15
Lysine	65.39
Methionine	12.26
Phenylalanine	26.61
Valine	41.36

Source: Data from Ehigiator and Oterai (2012).

Macrobrachium macrobrachion Herklots 1851 (= *Palaemon africanus*)

Order: Decapoda

Family: Palaemonidae

Common name: Brackish river prawn.

Distribution: North America and West Africa (from Senegal to northern Angola) (De Grave, 2013a).

Habitat: Freshwater and brackish freshwater.

Description: The rostrum of adult males is equal to or slightly longer than the antennal scale, straight or with tip curved slightly or occasionally strongly upward; often there is an untoothed portion near the tip, followed by one or two apical teeth. Second chelipeds are with carpus, which is slightly longer than the palm, and the palm is much longer than the fingers, which are straight and covered with a fur-like dense layer of short soft hairs. Chela, carpus, and merus are uniformly dark colored with a row of visible spines along the inner margin. Ischium is pale colored. Body is dark, with dorsal parts of the last three or four abdominal somites, which are light colored. The side of the carapace has a dark line running from below the eye toward the base of the second cheliped (Powell, 1983).

Nutritional Facts

Proximate Composition of Edible Portion (% DM)

	Dry Matter	Moisture	Ash	Protein	Fiber
M. vollenhovenii	94.13	5.87	20.00	53.85	1.00
M. macrobrachion	92.37	7.63	21.00	58.92	1.00

Source: Data from Ehigiator and Nwangwu (2011).

CRAYFISH (FRESHWATER LOBSTERS)

Astacus astacus Linnaeus 1758

Phylum: Arthropoda

Class: Crustacea

Order: Decapoda

Family: Astacidae

Common name: Noble crayfish, red-footed crayfish, red-clawed crayfish, broad-fingered crayfish.

Distribution: Throughout Europe (Edsman et al., 2010).

Habitat: Rivers, lakes, ponds, and reservoirs.

Description: This species varies in color from green or blue to brown and sometimes black. The undersides of the claws are dark red. The six segments of the abdomen are individually encased with a flexible membrane. It has a pair of large claws at the front end, followed by four pairs of walking legs and then four pairs of small swimming legs called swimmerets. A central tail flap is surrounded by four other flaps. There are two eyes on the end of eyestalks. The maximum lengths of male and female are 16 and 12 cm, respectively (ARKive).

Nutritional Facts

Proximate Composition (% DM)

Protein	Fat	Ash	Moisture[a]	Energy[b]
39.3	3.0	33.5	73.6	12.3

[a] Wet weight.

[b] kJ/g DW.

Source: Data from Huner et al. (1996).

Astacus leptodactylus Eschscholtz 1823

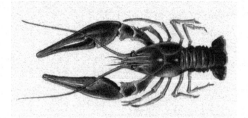

Order: Decapoda

Family: Astacidae

Common name: Turkish crayfish, narrow-clawed crayfish.

Distribution: Throughout Europe, eastern Russia, and the Middle East (Gherardi and Souty-Grosset, 2010).

Habitat: Fresh and brackish waters; rivers, lakes, canals, lagoons, estuaries.

Description: This species can grow up to 30 cm in length. The sides of the thorax are very rough, usually pale yellow to pale green in color. It has two pairs of post-orbital ridges, the second of which may have spines. It also has a prominent tubercle (small nodule) on the shoulder of the carapace. The claws are long and narrow. Their upper surface is rough and the underside is the same color as the body. A tubercle can be found on the fixed side of the claw. Relatively thinner fingers are seen on the claws.

Nutritional Facts

Proximate Composition of Whole Body Muscle (% WW)

Moisture	Protein	Lipid	Ash
68	19.82	6.31	8.78

Source: Data from Noverian et al. (2011).

Proximate Composition in Tail Muscle (%)

Moisture	Protein	Fat	Ash	Cholesterol[a]
78.19	14.61	0.57	1.39	81.51

[a] mg/100 g.

Fatty Acids in Tail Muscle (% of Total Fatty Acids)

SFA	MUFA	PUFA	n-3 PUFA	n-6 PUFA
25.56	28.17	45.97	32.74	13.23

Fat-Soluble Vitamins in the Tail Muscle (µg/100 g)

A	D2	D3	E[a]	K
0.43	0.54	0.26	1.19	0.76

[a] mg/100 g.

Source: Data from Harlioglue et al. (2012).

Procambarus clarkii Girard 1852

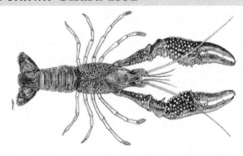

Order: Decapoda

Family: Cambaridae

Common name: Red swamp crayfish.

Distribution: From northern Mexico to Escambia County in Florida, north to southern Illinois and Ohio (Crandall, 2010a).

Habitat: Wet meadows, subterranean karst systems, seasonal swamps and marshes, permanent lakes and streams, rice paddy fields, and irrigation channels and reservoirs.

Description: Body shape of this species is cylindrical. Cephalothorax is conspicuously granular (roughened) in adults, provided with numerous small tuberculi, and also having strong cervical, cephalic, branchiostegal, and marginal spines. Rostrum is long with margins that are straight and convergent. Chelae are narrow and long, notched in proximal portion of the dactyl, leaving a gap and delimited by a tubercle. Large tubercle is seen at the opposite end of the gap on fixed finger. Large scarlet tubercles are seen on the palm and fingers. Color in adults is dark red, some in shades of brown. A wedge-shaped black stripe is present on the abdomen. Chelae are with bright red tubercles (McAlain and Romaire, 2007).

Nutritional Facts

Proximate Composition of Edible Portion (g/100 g DM)

	Moisture	Protein	Fat	Ash	CHO[a]	Energy[b]
Female	5.6	58.6	1.7	8.6	25.5	352
Male	7.0	62.6	3.1	10.2	17.5	374

[a] Carbohydrate.
[b] kcal.

Minerals in Muscle (mg/100 g)

	Ca	P	Fe	Zn	Se
Male	2843	343.6	8.9	15.1	0.9
Female	1474	327.2	11.7	9.5	1.4

Source: Data from Mona et al. (2000).

Essential Amino Acids (mg/100 g)

Lysine	7.41
Threonine	4.49
Cystine	1.19
Valine	4.71
Methionine	2.97
Isoleucine	4.27
Leucine	7.30
Tyrosine	3.54
Phenylalanine	4.56

Nonessential Amino Acids (mg/100 g)

Histidine	1.82
Arginine	7.83
Aspartic acid	9.72
Serine	3.99
Glutamic acid	3.61
Proline	3.99
Glycine	7.78
Alanine	4.25

Source: Data from Ahmad et al. (2013).

Fatty Acids (% of Total Fatty Acids)

Lauric C12:0	0.531
Myristic C14:0	0.726
Palmitic C16:0	16.53
Palmitaletic C16:1	2.13
Stearic C18:0	5.005
Oleaic C18:1	44.01
Linoleaic C18:2	12.5
Linolenic C18:3	2.58
Arachidic C20:0	4.59
Behenic C22:0	0.913
Erucic C22:1	10.42
Monioneic fatty acids (C16:1, C18:1, C22:1)	56.56
Polyioneic fatty acids (C18:2, C18:3)	15.08
Saturated fatty acid	26.311
Unsaturated fatty acid	73.689

Source: Data from Zaglol and Eltadawy (2009).

Procambarus zonangulus Hobbs & Hobbs 1990

Order: Decapoda

Family: Cambaridae

Common name: Southern white river crayfish.

Distribution: United States (Alabama, Louisiana, Maryland) (Crandall, 2010b).

Habitat: Permanent streams and aquaculture ponds.

Description: These crayfish have a space called an areola separating the sides of the back, forming a gap in the middle. Color of the animal is usually brown, with pink or purple in some adults. Mature crawfish have more elongated and cylindrical claws. Usually they have white or tan walking legs (Maryland Department of Natural Resources).

Nutritional Facts

Proximate Composition and Minerals (% DM)

Protein	Fat	Fiber	Ash	Moisture[a]	Ca	P	Energy[b]
39.9	4.8	11.5	32.8	72.0	13.0	0.54	14.2

[a] Wet weight.

[b] kJ/g.

Source: Data from Huner et al. (1996).

Orconectes rusticus Giard 1852

Order: Decapoda

Family: Cambaridae

Common name: Rusty crayfish.

Distribution: Native range is in the Ohio River and its tributaries; lakes

and rivers throughout Ohio, parts of Indiana and Kentucky (Phillips, 2010).

Habitat: Streams, lakes, and ponds with varying substrates from silt to rock and plenty of debris for cover.

Description: This species is easily distinguished from similar species by the rusty spots on each side of its carapace, which is tan in color. A rust-colored band is also seen down the center of the back side of the abdomen. Tips of claws have black bands. Oval gap is noticeable on claws when closed. Maximum size of the species is only 10 cm (http://www.seagrant.umn.edu/exotics/rusty.html).

Nutritional Facts

Proximate Composition and Minerals (% DM)

Protein	Fat	Fiber	Ash	Moisture[a]	Ca	P	Energy[b]
39.9	4.8	11.5	32.8	72.0	13.0	0.54	14.2

[a] Moisture.

[b] kJ/g.

Source: Data from Huner et al. (1996).

Orconectes limosus Rafinesque 1817

Order: Decapoda

Family: Cambaridae

Common name: Spiny-cheek crayfish.

Distribution: North America (Adams et al., 2010a).

Habitat: Clear streams that are 10–100 m wide, with silt, cobble, gravel, and sand substrates.

Description: This species is a small to medium-sized crayfish; the largest specimens reach just over 11 cm long. They have distinctive spiny cheeks, legs with orange tips, and striped abdomens, but are often colored black from the sediment in which they live (NNSS, 2014).

Nutritional Facts

Proximate Composition of Meat (% WW)

Water	Total N2	Protein N2
80.9	2.75	83

Source: Data from Dabrowski et al. (1966).

Proximate Composition of Meat (% WW)

Fat	Water
0.444	83.64

Fatty Acids (% of Total Fatty Acids)

SFA	MUFA	PUFA	PUFA n-3	PUFA n-6
32.89	34.35	32.76	13.70	19.07

Source: Data from Stanek et al. (2010).

Minerals (mg/kg WW)

Mn	Fe	Zn	Cu
0.26–91.3	0.54–81.1	6.80–51.91	1.21–4.34

Source: Data from Protasowicki et al. (2013).

Cherax destructor Clark 1936

Order: Decapoda

Family: Parastacidae

Common name: Yabbie crayfish.

Distribution: Throughout Australia (ARKive).

Habitat: Low-lying swamp grounds, streams, rivers, and dams.

Description: This smooth-shelled species usually varies in color from olive-green to brown, but can also be blue, yellow, red, or black depending on the habitat, location, and individual (http://www.arkive.org/yabbie-crayfish/ cherax-destructor/-ref3). The six segments of the abdomen are individually encased within a flexible membrane. It has a pair of large claws at the front end, followed by four pairs of walking legs and then four pairs of small swimming legs called swimmerets. A central tail flap is surrounded by four other flaps that are used to move the crayfish rapidly through the water. There are two eyes on the end of eyestalks. The maximum length and weight of this species are 20 cm and 320 g, respectively (ARKive).

Cherax tenuimanus Smith 1912

Order: Decapoda

Family: Parastacidae

Common name: Hairy marron, Margaret River marron.

Distribution: Southwestern Australia (ARKive).

Habitat: Permanent freshwater tributaries of forested high-rainfall areas.

Description: It is one of the largest freshwater crayfish species (2 kg) in the world. This hairy-shelled species has jet black pincers and a paler olive-green to brown body. Its underside is brown and females have areas of red coloration on the underside and some splashes of purple. The six segments of the abdomen are individually encased within a flexible membrane. It has a pair of large pincers at the front end, followed by four pairs of walking legs and then four pairs of small swimming legs called swimmerets. A central tail flap is surrounded by four other flaps. There are two eyes on the end of eyestalks (ARKive).

Nutritional Facts

Proximate Composition, Minerals, and Energy (% DM)

	CDW	CDS	CTW	CTS
Protein	47.3	38.7	46.5	45.8
Fat	3.6	7.5	5.1	7.3
Fiber	10.2	10.3	13.6	11.5
Ash	28.0	31.4	26.9	24.5
Moisture[a]	—	74.5	—	75.0
Ca	9.4	11.9	8.7	9.6
Energy[b]	14.6	16.3	12.3	18.7

CDW, *Cherax destructor* of Western Australia; CDS, *Cherax destructor* of South Australia; CTW, *Cherax tenuimanus* of Western Australia; CTS, *Cherax tenuimanus* of South Australia.

[a] Wet weight.

[b] kJ/g DM.

Source: Data from Huner et al. (1996).

Cherax quadricarinatus Von Martens 1868

Order: Decapoda

Family: Parastacidae

Common name: Australian red claw crayfish.

Distribution: Tropical Queensland, the Northern Territory, and southeastern Papua New Guinea; South Africa, Mexico, Jamaica, and Puerto Rico (Austin et al., 2010).

Habitat: Upper reaches of rivers of high turbidity, slow-moving streams, or static water holes.

Description: It is a relatively large freshwater crayfish with a smooth lustrous deep blue to green shell. Males exhibit bright red coloration on the margins of their large claws. Males can reach a maximum weight of 600 g and females 400 g. It is distinguished from other crayfish by size, color, and the presence of four distinct anterior ridges (carinae) of the carapace (Jones, 2011; NSW Department of Primary Industries).

Nutritional Facts

Proximate Composition[a] (g/100 g WW)

Dry Matter	Protein	Ash
24.36	23.7	2.07

[a] Experiment, fed with 32% protein diet.

Source: Data from Pavasovic et al. (2007).

Proximate Composition of Eggs (% WW)

Protein	Lipid	CHO[a]
63.2	32.3	4.4

[a] Carbohydrate.

Source: Data from Garcia-Guerro et al. (2000).

CRABS

Potamon potamios Olivier 1804

Phylum: Arthropoda

Class: Crustacea

Order: Decapoda

Family: Potamidae

Common name: Freshwater crab, semiterrestrial crab.

Distribution: Eastern Mediterranean, including many Mediterranean islands, extending as far south and west as the Sinai Peninsula (Cumberlidge, 2008a).

Habitat: Streams, rivers, and lakes.

Description: In this species, epigastric crests are only a little in advance of post-orbitals and are parallel to them. Post-orbitals are running in a straight line at right angles to the long axis of the carapace, hardly interrupted, or angulate at the anterior end of the cervical groove. The eighth thoracic sternite is incompletely separated by a longitudinal median line, and fused anteriorly at the suture between sternites 7 and 8 by a narrow transverse ridge interrupting the longitudinal line. Unlike other brachyuran crabs, the species of this family have no larval stages in their life history. Females lay only a few large eggs, which hatch directly into young crabs (Annandale and Kemp, 2013).

Nutritional Facts

Proximate Composition of Meat (% DM)

	Moisture	Protein	Fat	Ash
Female	78.85	13.94	0.67	1.05
Male	81.03	13.96	0.67	0.99

Mineral Contents of Meat (mg/g)

	Ca	Mg	K	Na	P	Fe	Cu	Zn
Female	13.07	1.74	8.93	11.58	0.76	0.031	0.06	0.20
Male	13.53	2.08	11.15	11.59	0.75	0.02	0.06	0.21

Source: Data from Bilgin and Fidanbaş (2011).

Paratelphusa masoniana Henderson 1893

Order: Decapoda

Family: Potaminidae

Common name: Freshwater crab.

Distribution: India (Manhas et al., 2013).

Habitat: Rivers, freshwater systems.

Description: The maximum carapace width and weight recorded are 6 cm and 85 g, respectively. The post-orbital crests are prominent. The abdomen of the adult male is regularly triangular. The sixth segment is never broad, its length almost always being equal to and seldom exceeding its distal breadth. The telson is never broadly triangular but is broadly semielliptical or tongue shaped, or at least elongate (Bahir and Yeo, 2007).

Nutritional Facts

Proximate Composition (% DM)

Protein	Lipid	Ash	Moisture[a]
54.28	4.83	8.37	80.98

[a] Wet weight.

Source: Data from Langer et al. (2013).

Sudanonautes africanus A. Milne-Edwards 1869

 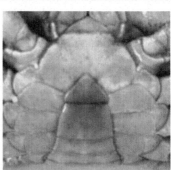

DORSAL AND VENTRAL SIDES

Order: Decapoda

Family: Potamonautidae

Common name: African river crab.

Distribution: Eastern part of West Africa; Nigeria, Cameroon, and Central Africa (Cumberlidge, 2008c).

Habitat: Streams, rivers, rain forests.

Description: In this species, the terminal segment of gonopod is thin and needle-like. Carapace is flat. Post-frontal crest meets the anterolateral margin at the epibranchial tooth. Carapace has patches of raised warts. Proximal region of pollex of propodus of the major cheliped has a large, flattened, fused tooth (Cumberlidge, 1994).

Nutritional Facts

Proximate Composition of Flesh (g/100 g DM)

	Protein	Fat	Ash	Fiber	Moisture	CHO[a]	Energy[b]
Male	34.50	3.25	7.36	4.88	9.41	45.40	1480
Female	17.30	3.11	10.10	1.86	8.54	59.10	1415

[a] Carbohydrate.
[b] kcal.

Minerals of Flesh (mg/100 g DM)

	Na	K	Ca	P	Mg	Zn	Fe	Mn	Co	Ni
Male	29.6	34.5	24.4	168	30.1	8.14	11.40	1.15	0.27	6.06
Female	34.3	36.5	30.9	153	30.0	7.90	15.0	0.87	0.50	7.46

Source: Data from Adeyeye et al. (2010).

Amino Acids (mg/g protein DM)

	Whole Body	Flesh
Lysine	58.0	50.0
Histidine	20.2	20.1
Arginine	40.1	51.3
Aspartic acid	59.0	68.5
Threonine	30.0	37.4
Serine	20.0	29.5
Glutamic acid	130.0	128.0
Proline	32.6	29.4
Glycine	41.1	61.7
Alanine	33.0	28.9
Methionine	21.1	22.8
Cysteine	11.0	10.2
Valine	29.0	37.9
Isoleucine	28.0	39.0
Leucine	60.9	60.0
Phenylalanine	40.5	38.1
Tyrosine	29.5	36.4
Protein[a]	32.5	24.8

[a] g/100 g.
Source: Data from Adeyeye and Kenni (2008).

Callinectes pallidus De Rocheburne 1883

Order: Decapoda

Family: Portunidae

Common name: Gladiator swim crab.

Distribution: Eastern Atlantic: From Mauritania to Angola (http://www.sealifebase.fisheries.ubc.ca/summary/Callinectes-pallidus.html).

Habitat: Sandy and muddy bottoms in brackish and marine waters at less than 30 m depth.

Description: Carapace of this species is rhombic with broad anterior and posterior edges and a rough surface. Sides of the carapace are elongated to form a lateral spine and anterolateral teeth are present. In addition, the fourth pair of walking legs in this species is broad and flattened to form a paddle, which helps in swimming. This species has a maximum carapace width of 10.9 cm (Akin-Oriola et al., 2005).

Cardisoma armatum Herklots 1851

Order: Decapoda

Family: Gecarcinidae

Common name: African rainbow crab.

Distribution: West Africa.

Habitat: River and coastal areas.

Description: Crabs of this species have a blue to purple carapace and orange to red legs. Carapace is apple shaped and smooth with no anterolateral teeth. Males are more colorful than females. This species has a maximum carapace width of 20 cm and has an average life span of 2–3 years (Akin-Oriola et al., 2005).

Nutritional Facts

Proximate Composition of Meat (% DM)

	Moisture	Protein	Fat	CHO[a]	Ash	Fiber
C. pallidus	53.56	24.38	2.09	5.40	13.41	1.16
C. armatum	51.50	23.94	1.65	7.13	14.96	1.14

[a] Carbohydrate.

Minerals of Meat (mg/kg DM)

	Ca	K	Mg	Na	Fe	Mn	Zn	Cu
C. pallidus	3844	1489	3843	2056	9.3	34.4	5.8	84.8
C. armatum	18,902	5720	26,462	1147	10.5	9.3	5.8	84.8

Source: Data from Elegbede and Fashina-Bombata (2013).

Spiralothelphusa hydrodroma Herbst 1794

Order: Decapoda

Family: Gecarcinucidae

Common name: Rice field crab.

Distribution: Northern Sri Lanka and Southeast India (Cumberlidge, 2008b).

Habitat: Freshwater, rice fields.

Description: Carapace of this species is transverse. Surfaces are smooth and convex. Epigastric and post-orbital cristae are distinct. Epibranchial teeth are well developed and distinct. Third maxilliped exopod has well-developed flagellum reaching beyond the width of the merus. Male's abdomen is broadly triangular. Male first pleopod is stout. Male second pleopod has distinct distal segment that is shorter than the elongate basal segment (Ng, 1994).

Nutritional Facts

Proximate Composition (% WW)

	Cephalothorax	Swimming/ Walking Legs
Protein	13.48	3.15
Carbohydrate	0.85	0.34
Lipid	0.57	0.19
Moisture	63.10	15.21
Ash	0.72	0.31

109

Mineral Composition (% WW)

	Cephalothorax	Swimming/Walking Legs
Calcium (Ca)	9.70	4.30
Magnesium (Mg)	0.81	0.36
Potassium (K)	5.20	3.40

Sodium (Na)	7.51	4.35
Iron (Fe)	0.01	0.08
Copper (Cu)	0.02	0.01
Zinc (Zn)	0.15	0.80

Source: Data from Varadharajan and Soundarapandian (2014).

Travancoriana schirnerae Bott 1969

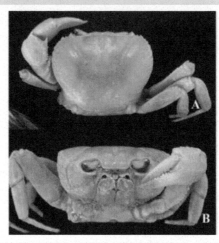

(A) DORSAL VIEW (B) FRONTAL VIEW

Order: Decapoda

Family: Gecarcinucidae

Common name: Freshwater crab.

Distribution: India (Tamil Nadu) (Cumberlidge, 2008d).

Habitat: Shallow (<30 cm deep), steep, rocky streams in shaded areas.

Description: Carapace of this species is broader than long. Dorsal surface is slightly convex in the frontal view, flat posteriorly, and smooth. Epistomal median lobe is without a median tooth. Post-orbital cristae are confluent with epigastric cristae. External orbital angle is broadly triangular, with long outer margin that is three to five times the length of the inner margin. Epibranchial tooth is small and blunt. Third maxilliped exopod has a long flagellum. Male abdomen is T-shaped to narrowly triangular in large males (Bahir and Yeo, 2007).

Nutritional Facts

Proximate Composition

	Protein (%)	Lipid (mg/100 g)	Moisture (%)	FAA (mg/100 g)	Cholesterol (mg/100 g)
Male	18.84	386.5	84.29	1197	18.0
Female	19.95	352.5	81.33	1794	23.6

FAA, free amino acid.

Source: Data from Devi and Smija (2013).

Eriocheir sinensis H. Milne-Edwards 1853

Order: Decapoda

Family: Varunidae

Common names: Chinese mitten crab, big sluice crab.

Distribution: Native to eastern Asia from Korea in the north to the Fujian Province of China in the south; Europe and North America.

Habitat: Freshwater, coastal estuaries, and salt water; upstream from the mouth.

Description: This species has a square-shaped carapace that is little longer than wide and markedly convex and uneven with four sharply edged epigastric lobes. Propodus of the fifth pereiopod is rather narrow and slender with dactylus that is claw shaped. Males produce an intensive mitten-like covering on the claws, with hairs at both inner and outer surfaces or only at the outer surface (always naked and smooth in other *Eriocheir* species). No longitudinal row of hairs is seen on the dorsal part of the propodus in the anterior two pairs of ambulatory legs. Front with four acuminated and deeply separated teeth. Four anterolateral teeth are distinct (not rudimentary) (Weimin, 2014).

Nutritional Facts

Proximate Composition of Meat (% WW)

Moisture	Protein	Fat	Ash
78.8	18.9	0.9	1.39

Minerals (mg/100 g)

Zn	Fe	K	Na	Mn	Cu	Mg	Ca	P
9.1	3.9	273	190	0.09	1.6	22	67	514

Amino Acids (g/100 g)

Aspartic acid	1.79
Glutamic acid	2.74
Serine	0.75
Histidine	0.43
Glycine	1.16
Threonine	0.86
Alanine	1.25
Arginine	1.78
Tyrosine	0.63
Cysteine	0.18
Valine	0.86
Methionine	0.23
Tryptophan	0.51
Phenylalanine	0.78
Isoleucine	0.78
Leucine	1.39
Lysine	1.46
Proline	0.52
Total	18.11

Fatty Acids (% of Total Fatty Acids)

SFA	MUFA	PUFA	Others	PUFA n-3	PUFA n-6
24.91	49.81	23.87	1.40	7.44	16.43

Source: Data from Chen et al. (2007).

111

Ozeotelphusa senex senex Fabricius (= *Oziotelphusa wagrakarowensis*)

Order: Decapoda

Family: Parathelphusidae

Common name: Freshwater edible crab.

Distribution: Sri Lanka and southern India (Bahir and Yeo, 2007).

Habitat: From the bank of shallow (<30 cm), slow-flowing, rocky streams.

Description: Dorsal surface of the carapace is highly convex. Anterior lateral carapace is low in the frontal view. Epistomal median lobe has distinct, sharp, pointed tooth. Frontal margin is bilobed in the dorsal view. Epibranchial tooth is moderate in size and sharp. Post-orbital region is concave. Post-orbital cristae are sharp and almost straight to curved. Branchial region is gently inflated. Male abdomen is triangular, with concave lateral borders. Segment 6 is trapezoidal, wider than long, and slightly longer than the telson, and has distinctly concave lateral borders, which form an hourglass shape (Bahir and Yeo, 2007).

Nutritional Facts

Proximate Composition (mg/g WW)

CHO[a]	Glycogen	Protein	FAA[b]	Lipid	FFA[c]
8.5	1.8	60.3	7.2	32.6	16.6

[a] Carbohydrate.

[c] Free amino acid.

[c] Free fatty acids.

Source: Data from Janakiram et al. (1983).

Habitat: Large river systems.

Description: In this species the head, eyes, and byssus are absent. The shell measures 70–200 mm in dimensions. The shell is oblong, with the margin opposite the hinge more or less straight or only slightly incurved. The margin opposite the hinge is outwardly curved. The shell is opaque (and robust), brown, bright or dark greenish or greenish brown and glossy (Dey, 2007).

Nutritional Facts

Proximate Composition (g/g DM)

	Protein	Lipid	CHO[a]	Ash
L. jenkinsianus	0.483	0.0472	0.2033	0.2665
L. generosus	0.440	0.0672	0.2705	0.2223

[a] Carbohydrate.

Source: Data from Shafakatullah et al. (2013).

Lamellidens marginalis Lamarck 1819

Order: Unionoida

Family: Unionidae

Common name: Indian freshwater mussel.

Distribution: Lower and upper Gangetic Plains in India and Bangladesh, Sri Lanka, Myanmar (Madhyastha et al., 2010).

Habitat: Lentic and lotic freshwater habitats; ditches, ponds, lakes, rivers, and streams; prefers stagnant waters.

Description: This species is cephalic and bilaterally symmetrical. Umbo is present along the dorsal margin and is slightly anterior to the hinge. True siphons are absent, and instead, there are two or three openings present in the mantle along the posterior margin. Adult unionids measure from 30 to 250 mm in length, and are just as variable in shape and color (Dey, 2007).

Nutritional Facts

Proximate Composition (% WW)

Moisture	Ash	Protein	Fat	CHO[a]	Fiber
85.9	2.184	6.464	0.507	4.943	0.0025

[a] Carbohydrate.

Minerals (mg/100 g DM)

Ca	P	Fe	Na	K
210.219	62.037	94.333	39.478	32.428

Source: Data from Baby et al. (2010).

Proximate Composition (g/g DM)

Protein	Lipid	CHO[a]	Ash
0.440	0.0459	0.2664	0.2477

[a] Carbohydrate.

Source: Data from Shafakatullah et al. (2013).

Parreysia corrugata Müller 1774

Order: Unionoida

Family: Unionidae

Common name: Freshwater mussel.

Distribution: India, Bangladesh, and Burma (Jadhav, 2009).

Habitat: Rivers, ponds, and lakes.

Description: This species is without a head but with a beak. Lateral teeth are posterior to the beak. It has a foot rather than a byssus. There is no true siphon. The inhalant aperture has branched papillae. This species attains sizes of 22, 36, 46, 51, 55, and 57 mm at the end of its first to sixth years, respectively. The life span of this mussel is about 6 years (Dey, 2007).

Nutritional Facts

Proximate Composition (DM)

Glycogen (mg/g)	Protein (%)	Lipid (%)
9.62	7.73	6.81

Minerals (mg/g)

Ca	Fe	Mn	Na	K	Cu	Zn	Mg
11.78	6.33	9.94	3.33	4.56	0.06–0.16	0.37–0.55	0.98–2.36

Source: Data from Malathi and Thippeswamy (2013).

Parreysia flavidens Benson 1862

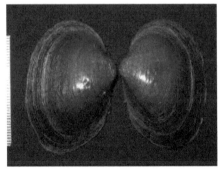

Order: Unionoida

Family: Unionidae

Common name: Freshwater mussel.

Distribution: Bangladesh, India.

116

Habitat: Lowland streams and rivers.

Description: Shells of this species are rounded, oval to elliptical, broad in proportions, thick, and short in contrast to Lamellidens. Cardinal teeth are narrowed and umbos are elevated with a strong ridge (Prabhakar and Roy, 2008).

Parreysia khadakvaslaensis **Ray 1966 (*Radiatula khadakvaslaensis*)**

Order: Unionoida

Family: Unionidae

Common name: Freshwater mussel.

Distribution: Worldwide (www.Animals.com).

Habitat: Reservoirs.

Description: This species is without a head but with a beak. Lateral teeth are posterior to the beak. It has a foot rather than a byssus. There is no true siphon. The inhalant aperture has branched papillae. Adult reaches a size range of 30–250 mm (Dey, 2007).

Nutritional Facts

Proximate Composition (% DM)

	Protein	Lipid	CHO[a]
P. favidens	41.2–60.8	3.8–8.2	14.79–42.3
P. khadakvaslaensis	40.6–57.2	3.2–7.6	18.3–40.2

[a] Carbohydrate.

Source: Data from Shetty et al. (2013).

Unio terminalis **Bourguignat 1852 (= *Unio crassus*)**

Order: Unionoida

Family: Unionidae

Common name: Common mussel.

Distribution: Jordan, Syria, Lebanon, and Israel (Lopes-Lima and Seddon, 2014a).

Habitat: River basin and coastal plain rivers; most suitable habitat is low-flow, muddy substrates.

Description: Color of the shell is brown to dark black. Cardinal tooth of the right valve is irregularly serrated. Lateral teeth are present. Green beams are frequently present on the outer side of valves. Length and height of the shell vary from 50 to 70 mm and 30 to 35 mm, respectively (http://inpn.mnhn.fr/docs/cahab/fiches/1032.pdf).

117

Potomida littoralis Cuvier 1798 (= *Unio littoralis*)

Order: Unionoida

Family: Unionidae

Common name: Freshwater mussel.

Distribution: South Europe (Portugal, Spain, France, Greece), North Africa (Tunisia, Algeria, Morocco), Middle East (Armenia, East Turkey, Syria, Palestine, Israel) (Schultres, 2012).

Habitat: Rivers, river canals, slow-moving waters, river affluents, canals, lakes.

Description: Shell of this species is greenish black and interior with slightly violet hue. Shell shape is variable but usually short and high. Embryonic shell is usually white without periostracum. Teeth are very strong. Animal is yellowish or greenish gray, and its mantle is yellow. Gills are dark reddish brown and foot is yellowish. Shell measures 40–50 × 50–70 × 25–30 (height) mm (Schultres, 2012).

Nutritional Facts

Proximate Composition (%)

	Moisture	Protein	Lipid	Ash
U. terminalis	80.36	11.87	2.55	1.68
P. littoralis	81.69	11.97	1.05	1.61

Fatty Acids (%)

	U. terminalis	*P. littoralis*
SFA	32.13	30.21
MUFA	19.60	22.84
PUFA	37.08	32.41

Source: Data from Ersoy and Şereflişan (2010).

Unio tigridis Bourguignat 1853

Order: Unionoida

Family: Unionidae

Common name: Freshwater bivalve.

Distribution: Turkey and Syria, Iraq, Iran (Lopes-Lima and Seddon, 2014b).

Habitat: Rivers and lakes.

Description: This species has a solid elongate shell usually with thick walls. It is sexually diocious, with relative female excesses most of the time. Length and width of the shells are 3.5 cm and 1.47 cm, respectively. Shells with larger size approaching 5 cm are also common (Al-Bassam and Hassann, 2006).

Nutritional Facts

Proximate Composition (Freeze-Dried)

Lipids (%)	Proteins (mg/g)
32.29	9.479

Source: Data from Salman and Nasar (2013).

CLASS: GASTROPODA (SNAILS)

Lissachatina fulica Bowdich 1822 (= *Achatina fulica*)

Phylum: Mollusca

Class: Gastropoda

Order: Stylommatophora

Family: Achatinidae

Common name: Giant African snail.

Distribution: Throughout East Africa, ranging from Mozambique in the south to Kenya and Somalia in the north; Morocco, the Ivory Coast, and Ghana; throughout parts of Asia, Caribbean, and Australia (ARKive).

Habitat: Forest margins, along the edges of streams and rivers.

Description: Adult snail is about 7 cm in height, 20 cm in length, and 32 g in weight. Shell has a conical shape, being about twice as high as it is broad. When fully grown, shell has seven to nine whorls. Either clockwise (dextral) or counterclockwise (sinistral) directions can be observed in the coiling of the shell, although the right-handed (dextral) cone is more common. Shell coloration is highly variable, and dependent on diet. Typically, brown is the predominant color and shell is banded.

Nutritional Facts

Proximate Composition (% WW)

Protein	Fat	Fiber	Ash
29.82	4.25	0.11	11.83

Minerals (mg/kg WW)

Na	K	Ca	P
0.163	0.375	0.183	0.339

Source: Data from Adebayo-Tayo et al. (2011).

119

Helix sp.

Order: Stylommatophora

Family: Helicidae

Common name: Common brown snail.

Distribution: Native to Europe and the regions around the Mediterranean Sea.

Habitat: Native to freshwater habitats; introduced throughout the world.

Description: The adult of this unidentified species bears a hard, thin calcareous shell of 25–40 mm in diameter and 25–35 mm in height, with four or five whorls. The shell is variable in color and shade but generally is dark brown, brownish golden, or chestnut with yellow stripes, flecks, or streaks (characteristically interrupted brown color bands). The aperture is large and characteristically oblique. The body is soft and slimy, brownish gray. It has no operculum; during dry or cold weather it seals the aperture of the shell with a thin membrane of dried mucus.

Nutritional Facts

Proximate Composition (% WW)

Moisture	Ash	Protein	Fat	CHO[a]	Fiber
83.2	1.036	8.640	0.571	6.548	0.0035

[a] Carbohydrate.

Minerals (mg/100 g DM)

Ca	P	Fe	Na	K
80.757	60.447	19.181	83.954	60.447

Source: Data from Baby et al. (2010).

Anisus convexiusculus Hutton 1849

Order: Basommatophora

Family: Planorbida

Common name: Ram's horn snail.

120

Distribution: Southeast England, Belgium to North Italy, Denmark, Estonia, Greece, Ukraine (AnimalBase).

Habitat: Lives on water plants in freshwater.

Description: Ram's horn snails usually have shells formed like a disc or a plate. The disc-shaped shell of a ram's horn snail usually is tilted to the left, so the snail can carry it. So it appears as if ram's horn snails are coiled to the right. The fact that genital opening and respiratory opening are on the left shows that ram's horn snails indeed are sinistral (coiled to the left) (Living World of Molluscs).

Nutritional Facts

Proximate Composition (% WW)

Moisture	Ash	Protein	Fat	CHO^a	Fiber
75.7	4.607	12.927	0.972	5.793	0.0370

^a Carbohydrate.

Minerals (mg/100 g DM)

Ca	P	Fe	Na	K
253.210	204.107	139.201	77.755	58.316

Source: Data from Baby et al. (2010).

Nucella lapillus Linnaeus 1758

Order: Neogastropoda

Family: Muricidae

Common name: Dog whelk, Atlantic dogwinkle.

Distribution: Throughout North Atlantic from the Arctic to the Algarve in the east, Iceland and the Faroes, and from Long Island north to Southwest Greenland in the west (BIOTIC).

Habitat: Rocky shores and estuarine conditions.

Description: Shell is small, broadly conical, bearing spiral ridges, and consisting of a short, pointed spire, dominated by the last whorl. Shell is usually up to 3 cm in height by 2 cm broad but may reach up to 6 cm in height. Overall shell shape varies quite widely according to the degree of exposure to wave action of the shore on which a particular population lives, but the body whorl is usually around three-fourths of the total length of the shell. Aperture is usually crenulated in mature dog whelks. The external shell color is usually a whitish gray, but can be a wide variety of orange, yellow, brown, black, or banded with any combination of these colors. They can even, occasionally, be green, blue, or pink (MarLIN).

Nutritional Facts

Proximate Composition of Foot Tissue (% DM)

Moisture[a]	Protein	Fiber	Ash
73.69	82.25	1.50	8.0

[a] Wet weight.

Minerals of Foot Tissue (mg/100 g)

P	K	Fe	Zn	Ca	Mg
60.19	72.40	11.09	1.32	181.49	31.73

Source: Data from Eneji et al. (2008).

Lanistes varicus **Müller 1774**

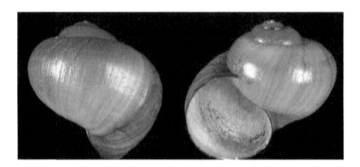

Common name: Apple snail.

Distribution: Africa and Madagascar.

Habitat: Irrigation channels, streams and lakes; inhabits shallow, muddy pools of river banks.

Description: Snails of this genus have a hyperstrophic sinistral shell, but the body of it is dextral. Size of the shell is 60 × 65 mm. Shell is approximately globose with evenly curved whorls.

There is a large aperture and wide umbilicus. Shell color is brown and unbranded (Brown, 2002).

Nutritional Facts

Proximate Composition of Foot Tissue (% DM)

Moisture[a]	Protein	Fiber	Ash
75.80	70.00	1.25	8.0

[a] Wet weight.

Minerals of Foot Tissue (mg/100 g)

P	K	Fe	Zn	Ca	Mg
62.52	69.39	10.10	1.30	152.70	30.56

Source: Data from Eneji et al. (2008).

122

Pila ampullacea **Linnaeus 1758**

Order: Architaenioglossa

Family: Ampullariidae

Common name: White apple snail.

Distribution: Southeast Asia (Sri-aroon and Richter, 2012)

Habitat: Paddy fields, irrigation canals, and similar habitats.

Description: Shell of this large Asiatic apple snail varies from 90 to 100 mm high, 85 to 90 mm wide. They have a globose shell with an oval shell opening (aperture). Spire is rather short and umbilicus is narrow to nearly closed. Surface of the shell is smooth. Color of shell varies from bright green to orange-brown with reddish spiral bands. The internal part of the shell is yellowish with a tinge of purple and marked with strong spiral bands, lighter at the lip. Operculum is two times higher than wide and calcified in older snails. Body is gray-brown (http://www.applesnail.net).

Nutritional Facts

Proximate Composition (%)

Moisture	Protein	Lipids	Ash	Fiber
76.32	10.67	0.06	5.54	0.03

Minerals (mg/100 g)

Ca	K	P	Fe	Na	Mg	Zn
129.18	71.13	60.52	10.19	0.04	31.19	1.31

Source: Data from Obande et al. (2013).

Pila globosa **Swainson 1822**

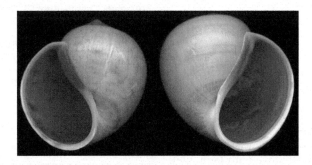

Common name: Apple snail.

Distribution: Nepal, southwestern Asia, Africa, North and South America, India (Budha et al., 2010b; Rao, 1989).

Habitat: Permanent and temporary stagnant freshwater bodies.

Description: The shell of this species is globose with an oval opening. It has a large and deep umbilicus. The color varies from olive green to gray green with a tinge of red. A large number of variations are known. The interior of the shell is dull reddish with very faint spiral bands visible, white at

the columella. The operculum is calcified at the inside (part attached to the snail) (Dey, 2007).

Nutritional Facts

Proximate Composition (% WW)

Moisture	Ash	Protein	Fat	CHO[a]	Fiber
85.5	2.599	8.272	0.725	2.902	0.0256

[a] Carbohydrate.

Minerals (mg/100 g DM)

Ca	P	Fe	Na	K
304.427	133.356	99.147	43.485	31.889

Source: Data from Baby et al. (2010).

Pila ovata Olivier 1808

Common name: Apple snail.

Distribution: Sudan, Mozambique, Nigeria, eastern Africa, Egypt (Ghamizi et al., 2010).

Habitat: Temporary freshwater pools, papyrus swamps, and stony beaches.

Description: The shell of this medium-sized (55–59 mm wide, 43–47 mm high) species has a relatively high spire (about half of the aperture height) and a round shell opening (aperture). The umbilicus is small but deep, and the lip somewhat thickened. The color varies from light brown to reddish olive with faint, non-continuous, spiral bands. Operculum

is 1.5 to 1.7 times higher than wide and calcified (Brown, 2002).

Nutritional Facts

Proximate Composition (% DM)

	Dry		
Protein	Matter	Ash	Fat
67.2–74.2	86.0–87.3	1.2–2.5	7.9–12.2

Source: Data from Chima and Akobundu (2010).

Minerals of Viscera (mg/g)

Cu	Pb	Fe	Zn
0.14	0.92	0.78	0.72

Source: Data from Aboho et al. (2009).

Pila virens Lamarck 1822

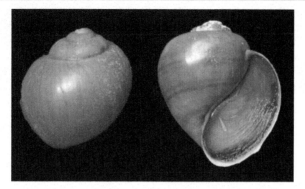

Common name: Apple snail.

Distribution: Southern India (Budha and Madhyastha, 2010).

Habitat: Lentic and lotic habitats such as streams, ponds, tanks, paddy fields.

Description: This species has a globose shell with a sharp spire. Height of the shell is from 40 to 65 mm. Shallow sutures (angle 90° and more) and medium-sized umbilicus are present. The shell surface is smooth in most cases, although more eroded forms are also seen. The color can vary from uniform yellow-olive to dark brown. Reddish brown spiral bands can be absent at the outside in some varieties. Operculum is corneous and calcified at the inside in older snails (Dey, 2007).

Nutritional Facts

Proximate Composition (mg/g WW)

CHO[a]	Glycogen	Protein	Free AA	Lipid	Free FA
11.6	3.1	49.4	4.8	29.6	19.7

[a] Carbohydrate.

Source: Data from Janakiram et al. (1983).

Lanistes libycus Morelet 1848

Order: Architaenioglossa

Family: Ampullariidae

Common name: Freshwater snail.

Distribution: West Africa (Jørgensen et al., 2010).

Habitat: Lakes, marshes, and streams.

125

Description: Shells of this species have a shell length of 26.5–43.5 mm. Shell is globose, yellowish brown with darker spiral bands. Its spire is much less than the height of the aperture. The presence of a gill, accessory lung sac inside the mantle cavity, and anterior kidney chamber is the characteristic feature that helps this species to adapt to an amphibian life (Brown, 2002; arnobrosi.tripod.com/snails/lanistes.html).

Pomacea bridgesii Reeve 1856

Order: Architaenioglossa

Family: Ampullariidae

Common name: Golden apple snail, common apple snail, golden mystery snail.

Distribution: Paraguay, Brazil, and Bolivia (Pastorino and Darrigan, 2011).

Habitat: Lakes and rivers.

Description: Shell of this species has five or six whorls. Most obvious characteristic features of the shell are the square shoulders (flat at the top of the whorls) and almost 90° sutures. Shell opening (aperture) is large and oval and the umbilicus is large and deep. Size of the shell varies from 40 to 50 mm wide and 45 to 65 mm high. Spire is high and sharp. They are the most colorful of all apple snail species, ranging in color from the wild brown, gold, albino, chestnut to blue, jade, and even shades of purple and burgundy, with or without stripes. Operculum is moderately thick and corneous. The color of the operculum varies from light to dark brown. The operculum can be retracted in the aperture (shell opening) (http://www.applesnail.net; http://www.petfish.net).

Pomacea canaliculata Lamarck 1819

Common name: Channeled apple snail.

Distribution: Tropical and subtropical South America.

Habitat: Freshwater habitats.

Description: Shell of this species is globose and relatively heavy (especially in older snails). Five to six whorls are separated by a deep, indented suture. Shell opening (aperture) is large and oval to round. Males are known to have a rounder aperture than females. Umbilicus is large and deep. Maximum size of these snails varies from 40 to 60 mm wide and 45 to 75 mm high depending on the conditions. Maximum length, however, is 150 mm. Normal coloration typically includes bands of brown, black, and yellowish tan; color patterns are extremely variable. Albino and gold color variations exist.

Nutritional Facts

Proximate Composition (% WW)

	Protein	Fat	Fiber	Ash
Lanistes libycus	31.66	3.57	0.09	12.21
Pomacea canaliculata	31.38	3.26	0.11	11.38
Pomacea bridgesii	33.24	3.18	0.13	11.42

Minerals (mg/kg WW)

	Na	K	Ca	P
Lanistes libycus	0.139	0.401	0.185	0.348
Pomacea canaliculata	0.152	0.412	0.196	0.314
Pomacea bridgesii	0.419	0.418	0.191	0.326

Source: Data from Adebayo-Tayo et al. (2011).

Tympanotonus fuscatus var. *radula* Linnaeus 1758

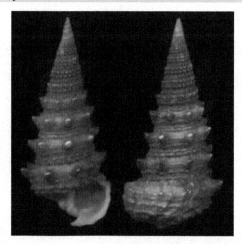

Order: Sorbeoconcha

Family: Potamididae

Common name: Mudflat periwinkle.

Distribution: Angola, Cape Verde, Gabon, and generally West Africa.

Habitat: Brackish waters of the riverine areas, mangrove swamps.

Description: This species is characterized by its turrented, granular, and spiny shells with tapering ends. Shell measures 80 × 25 mm. Its sculpture is of two forms: small tubercles or coarse ribs with short spines (Egonmwan, 2008).

Nutritional Facts

Proximate Composition (% DM)

Ash	Moisture	Protein	Fat	Fiber	CHO[a]
9.56	13.45	74.74	1.32	0.74	0.19

[a] Carbohydrate.

Minerals (mg/100 g)

Ca	Mg	Zn	Cu	Mn	Na	K	Fe
41.98	176.86	0.70	1.40	0.40	90.00	36.00	3.00

Amino Acids (mg/g protein)

Lysine	42.0
Histidine	20.0
Arginine	54.3
Aspartic acid	86.0
Threonine	33.2
Serine	35.5
Glutamic acid	134.8
Proline	30.7
Glycine	40.5
Alanine	38.5
Cystine	11.7
Valine	37.9
Methionine	11.7
Isoleucine	25.2
Leucine	87.5
Tyrosine	35.3
Phenylalanine	45.5

Source: Data from Ogungbenle and Omowole (2012).

Bellamya (Viviparus) bengalensis Lamarck 1882

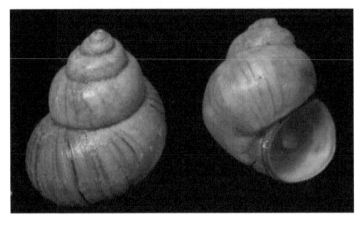

Order: Architaenioglossa

Family: Viviparidae

Common name: River snail.

Distribution: All over the southern Asia region; Iran, Bangladesh, Nepal, Pakistan, Myanmar, Sri Lanka, and throughout India (Budha et al., 2010a).

Habitat: Almost all types of lowland water bodies, mainly stagnant water and low-saline water resources such as rivers, streams, lakes, ponds, wetlands, marshes, ditches, paddy fields.

Description: Shell of this species is more or less oval and acuminate with variable and irregular dark bands. Adults of this species have a length of >25 mm and average weight of 3.5 g. They are livebearers; i.e., the juvenile snails are retained inside the mantle of the mother for several weeks. This species feeds primarily on particulate detritus in soft sediments (Brown, 2002).

Nutritional Facts

Proximate Composition (% WW)

Moisture	Ash	Protein	Fat	CHO[a]	Fiber
82.1	3.640	8.966	0.984	4.308	0.0355

[a] Carbohydrate.

Minerals (mg/100 g DW)

Ca	P	Fe	Na	K
166.404	128.849	100.717	53.687	39.370

Source: Data from Baby et al. (2010).

5
FISHES

ORDER: CYPRINIFORMES (CARPS)

FAMILY: CYPRINIDAE (MINNOWS OR CARPS)

Catla catla Hamilton 1822

Phylum: Chordata

Class: Actinopterygii

Order: Cypriniformes

Family: Cyprinidae

Common name: Catla.

Distribution: Asia: Pakistan, India, Bangladesh, Nepal, and Myanmar (Luna, 1988d).

Habitat: Freshwater and brackish water; rivers, lakes, and culture ponds.

Description: Body of this species is short and deep, somewhat laterally compressed. Head is very large, and its depth exceeding half the head length. Snout is bluntly rounded. Eyes are large and visible from underside of the head. Mouth is wide and upturned, with prominent protruding lower jaw. Upper lip is absent and lower lip is very thick. There are no barbels. Dorsal fin is inserted slightly in advance of pelvic fins. Anal fin is short and pectoral fins are long, extending to pelvic fins. Caudal fin is forked. Body is grayish on back and flanks, silvery white below, and fins are dusky. This species grows to a maximum size of 1.28 m and 45 kg (Jena, 2014; Natarajan and Jhingran, 1963).

Nutritional Facts

Proximate Composition (% WW)

Moisture	Protein	Fat	Ash
77.9	1.66	1.22	1.10

Source: Data from Babu et al. (2013).

Minerals (µg/kg)

Ca	Zn	Fe	K	Mn
8.1×104	93.8	1.9×103	238.6	234.9

Source: Data from Manirujjaman et al. (2014).

131

Labeo rohita F. Hamilton 1822

Order: Cypriniformes

Family: Cyprinidae

Common name: Rohu.

Distribution: South Asia: India, Bangladesh, Nepal, Myanmar, and Pakistan (Froese and Pauly, 2013).

Habitat: Rivers, ponds.

Description: Body of this species is bilaterally symmetrical and moderately elongated. Its dorsal profile is more arched than the ventral profile. Head is without scales and snout is fairly depressed, projecting beyond mouth. Eyes are dorsolateral in position, not visible from outside of head. Mouth is small and inferior. Lips are thick and fringed with a distinct inner fold to each lip. A pair of small maxillary barbels is concealed in a lateral groove. Dorsal fin is inserted midway between snout tip and base of caudal fin. Pectoral and pelvic fins are laterally inserted. Caudal fin is deeply forked. Color of body is bluish on back and silvery on flanks and belly.

It reaches a maximum length of 2 m and a weight of about 110 kg (Jena, 2014).

Nutritional Facts

Proximate Composition (g/100 g Edible Portion, WW)

Energy[a]	Water	Protein	Fat	CHO[b]	Ash
87	79.2	16.9	1.5	1.4	1.0

[a] kcal.

[b] Carbohydrate.

Minerals (mg/100 g Edible Portion)

Ca	P	Fe	Na	K
19	174	1.2	32	86

Vitamins (mg/100 g Edible Portion)

Retinol[a]	Riboflavin	Niacin
50	0.08	2.7

[a] µg.

Source: Data from Tee et al. (1989).

Amino Acids (% DM)

Methionine	Tryptophan
0.95	1.20

Source: Data from Ali et al. (2013).

Labeo pangusia Hamilton 1822

Order: Cypriniformes

Family: Cyprinidae

Common name: Pangusia labeo.

Distribution: Asia: Pakistan, India, Nepal and Bangladesh, Afghanistan, Bhutan, and Myanmar (Torres, 1991f).

Habitat: Rivers, lakes, and ponds; large streams.

Description: Body of this species is elongated and dorsal profile is more convex than that of ventral. Mouth is small, and it overhangs with distinct lateral lobes. Eyes are small and lips are thick and not fringed. Dorsal fin is inserted to snout tip. Pectorals do not extend up to pelvics. Caudal fin is deeply forked. Lateral line is complete, and there are two black spots just above and below the lateral line. Riverine fishes are brownish above and yellowish and white at sides and below.

Stream fishes, on the other hand, are dark or blackish above and light yellowish below. This species reaches a maximum size of 90 cm and 2.5 kg (Fahad, 2011e; Practical Fishkeeping).

Nutritional Facts

Proximate Composition (% DM)

Moisture	Protein	Lipid	Ash
12.20	70.08	8.487	5.65

Macroelements (mg/100 g)

Ca	Mg	K	Na	P
9.70	56.57	234.25	58.87	322.22

Microelements (mg/100 g)

Cu	Co	Mn	Zn	Fe	Ni	Cr
0.500	3.875	0.500	3.750	5.969	1.800	0.035

Source: Data from Hei and Sarojnalini (2012).

Labeo gonius **Hamilton 1822**

Order: Cypriniformes

Family: Cyprinidae

Common name: Kuria labeo.

Distribution: Asia: Pakistan, India, Bangladesh, Afghanistan, and Nepal (Shrestha, 1994).

Habitat: Lotic and lentic freshwater habitats.

Description: Body of this species is elongated and its dorsal profile is more convex than that of the ventral. Mouth is blunt, narrow, and subinferior. Lips are thick and fringed. Eyes are moderate in size. Two very short pairs of barbels are present. Dorsal fin is inserted. Pectoral fins are long. Caudal fin is deeply forked. Body is greenish black on its back. This species reaches a maximum size of 1.5 m and 1.4 kg (Fahad, 2011).

Nutritional Facts

Proximate Composition (% WW)

Moisture	Lipid	Protein	Ash	Fiber	CHO[a]
71.7	5.2	16.5	5.5	0.7	0.9

[a] Carbohydrate.

Source: Data from Begum et al. (2013).

Labeo boggut Sykes 1839

Order: Cypriniformes

Family: Cyprinidae

Common name: Boggut labeo.

Distribution: Asia: Pakistan, India, and Bangladesh (Bailly, 1997f).

Habitat: Upper reaches of rivers, ponds.

Description: Body of the species is elongated and slender with a more convex dorsal profile than the ventral profile. Snout and lips are thick. Mouth is moderate. There is one pair of short maxillary barbels. Pectoral fins are as as long as the head and do not extend to pelvic fins. Caudal fin is deeply forked. Body is silvery and darkest on back. Fins are orange in color. A dark spot is seen near the base of the caudal fin. This species reaches a maximum size of 29 cm and 350 g (Fahad, 2011d).

Nutritional Facts

Proximate Composition (per 100 g WW)

Protein (g)	Lipid (mg)	Moisture (%)
11.25	01.16	79.00

Source: Data from Patole (2012).

Labeo calbasu Hamilton 1822

Order: Cypriniformes

Family: Cyprinidae

Common name: Orangefin labeo.

Distribution: Asia: Pakistan, India, Bangladesh, Myanmar, Nepal, Thailand, and southwestern China (Reyes, 1993f).

Habitat: Freshwater and brackish water; rivers and ponds.

Description: Dorsal profile of this species is more convex than the abdomen. Lips are thick and fringed. Two pairs of barbels are present, and rostral pair is longer than maxillary pair. Caudal peduncle is short. Lateral line is well marked. Scales are of moderate size. Mouth is wide and inferior. Color of the body is dark black, but ventral is light-dark. Ventral portion of opercular region is white and iris is coppery. This grows to a maximum size of 90 cm and 8 kg (Bashar, 2010a; Reyes, 1993f).

Nutritional Facts

Proximate Composition (% WW)

Moisture	Ash	Protein	Lipid
79.80	1.08	16.47	2.65

Source: Data from Ahmed et al. (2012).

Labeo bata Hamilton 1822

Order: Cypriniformes

Family: Cyprinidae

Common name: Bata.

Distribution: Asia: India, Bangladesh, and Pakistan (Reyes, 1993e).

Habitat: Rivers.

Description: Body of this species is elongated. Its dorsal profile is more convex than the ventral. Snout projects slightly beyond the mouth and is often studded with pores. A pair of small maxillary barbels is hidden inside the labial fold. Dorsal fin originates midway between snout tip and anterior base of anal fin. Pelvics originate slightly nearer to snout tip than to caudal base. Body is bluish or darkish on upper half, silvery below, and the opercle is light orange. This species reaches a maximum size of 90 cm and 450 g.

Nutritional Facts

Proximate Composition (% WW)

Moisture	Ash	Protein	Lipid
79.48	1.36	15.42	3.73

Source: Data from Ahmed et al. (2012).

Cirrhinus mrigala Hamilton 1822

Order: Cypriniformes

Family: Cyprinidae

Common name: Mrigal carp.

Distribution: Asia: Native to large rivers in the Indian subcontinent; Pakistan, India, Nepal, and Bangladesh (Torres, 1991c).

Habitat: Fast-flowing streams and rivers.

Description: Body is bilaterally symmetrical and streamlined, and its depth is about equal to the length of the head. Head is without scales. Snout is blunt, often with pores. Mouth is broad and transverse. Upper lip is entire and not continuous with lower lip Lower lip is most indistinct. A single pair of short rostral barbels is seen. Origin of dorsal fin is nearer to end of snout than base of caudal. Pectoral fins are shorter than head, and caudal fin is deeply forked. Anal fin is not extending to caudal fin. Body is dark gray above and silvery below. Dorsal fin is grayish, and pectoral, pelvic, and anal fins are orange-tipped (especially during breeding season). This species attains a maximum size of 99 cm and 12.7 kg (Torres, 1991c; Ayyappan, 2014).

Nutritional Facts

Proximate Composition (% WW)

Moisture	Protein	Fat	Ash	CHO[a]
78.6	15.6	1.62	1.55	1.50

[a] Carbohydrate.

Source: Data from Babu et al. (2013).

Cirrhinus cirrhosus Bloch 1795

Order: Cypriniformes

Family: Cyprinidae

Common name: White carp, mrigal carp.

Distribution: Indian subcontinent (Luna, 1988e).

Habitat: Fast-flowing streams and rivers.

Description: Body of this species is elongated and torpedo shaped. Dorsal profile is more convex than ventral profile. Body is grayish or greenish on back and silvery on sides and abdomen. Fins are slightly orange. Lateral line is complete. It grows to a maximum size of 1 m and 12.7 kg (Galib, 2010b).

Nutritional Facts

Proximate Composition (Sundried Flesh g/100 g)

Moisture	Ash	Lipid	Protein
75.82	1.50	0.95	15.22

Minerals (Sundried Flesh µg/kg)

Ca	Zn	Fe	K	Mn
241.5	97.0	2.1×10^3	246.1	212.0

Source: Data from Manirujjaman et al. (2014).

Cyprinus carpio Linnaeus 1758

Order: Cypriniformes

Family: Cyprinidae

Common name: Wild common carp.

Distribution: Europe and Asia (Torres, 1991e).

Habitat: Warm, deep, slow-flowing and still waters, such as lowland rivers and large, well-vegetated lakes.

Description: Body of this species is elongated and somewhat compressed. Lips are thick. Two pairs of barbels are seen at angle of the mouth. Shorter ones are on the upper lip. Dorsal fin outline is concave anteriorly. Lateral line has 32–38 scales. Color of body is variable. Wild carps are brownish green on the back and upper sides, shading to golden yellow ventrally. Fins are dusky, ventrally with a reddish tinge (Peteri, 2014).

Nutritional Facts

Proximate Composition (% WW)

Protein	Moisture	Lipid	Ash
17.48	79.55	2.25	1.17

Fatty Acids (% of Total Fatty Acids)

SFA	MUFA	PUFA	PUFA n-3	PUFA n-6
28.47	38.57	32.52	5.13	24.98

SFA, saturated fatty acid; MUFA, monounsaturated fatty acid; PUFA, polyunsaturated fatty acid.
Source: Data from Dejana et al. (2013).

Amino Acids (g/100 g Proteins)

Aspartic acid	9.60
Glutamic acid	14.40
Serine	4.40
Glicine	7.00
Treonine	4.70
Arginine	6.90
Alanine	7.20

Tyrosine	3.30
Valine	5.90
Phenylalanine	4.00
Isoleucine	4.60
Leucine	8.30
Lysine	8.20
Proline	4.80
Cystine	0.90
Methionine	3.40
Other amino acids[a]	2.40

[a] Estimated by subtracting the total amount of amino acids quantified from 100 g proteins.
Source: Data from Aprodu et al. (2012).

Protein Content of the Meat

% of Wet Weight	% of Dry Matter
16.1	54

Lipid Content of the Meat

% of Wet Weight	% of Dry Matter
12.7	42

Major Elements in the Muscle (mg/100 g)

Na	K	Mg	Ca	P
30	387	51	63	247

Trace Elements in the Muscle (g/100 g)

Fe	Mn	Se
700	55	0

Vitamins in the Muscle (g/100 g)

A	B1	B2	B6
44	68	53	150

Source: Data from Steffens et al. (2006).

Cyprinus carpio communis Linnaeus

Order: Cypriniformes

Family: Cyprinidae

Common name: Scale carp.

Distribution: Native species of temperate region of Asia, especially China; introduced as a cultivated carp species throughout the world (Mohsin, 2012).

Habitat: Ponds, rivers, lakes, canals, beels, baors, etc., both open and closed water bodies.

Description: In this variety of common carp, body is elongated and head is comparatively very small. Dorsal of body is very convex and abdomen is bulky. Body is compressed and snout is rounded. Whole body is covered with moderately sized, regular concentric scales. Abdomen is rounded. Dorsal side is brownish. It is an omnivorous bottom feeder on the larvae of insects, worms, and molluscs, and stalks and leaves of submerged plants and occasionally on zooplankton (Zakia, 2014; Santhanam et al., 1987).

Cyprinus carpio specularis Lacepede 1803

Order: Cypriniformes

Family: Cyprinidae

Common name: Mirror carp.

Distribution: Native carp of Europe; introduced to Bangladesh from Nepal for aquaculture purposes (Rahman, 2005).

Habitat: Rivers, floodplains, and culturable ponds.

Description: This variety is characterized by the presence of large, shiny, scattered scales running along the side of the body in several rows with the rest of the body naked. It is an omnivorous bottom feeder on the larvae of insects, worms, and molluscs, and stalks and leaves of submerged plants and occasionally on zooplankton (Chumchal, 2002; Santhanam et al., 1987).

Nutritional Facts

Proximate Composition (% WW)

	Moisture	Protein	Fat	Ash
C. carpio communis	76.14	16.60	3.58	2.10
C. carpio specularis	77.46	14.63	3.80	2.02

Source: Data from Ganai (2012).

Hypophthalmichthys molitrix Valenciennes 1844

Order: Cypriniformes

Family: Cyprinidae

Common name: Silver carp.

Distribution: Asia: Native to most major Pacific drainages of East Asia; China and Eastern Siberia. Introduced around the world for aquaculture (Luna, 1988k).

Habitat: Rivers with marked water-level fluctuations.

Description: In this species, barbels are absent. Keels extend from isthmus to anus. The edge of the last simple dorsal ray is not serrated. A sharp scaleless keel is seen from the pectoral region to the anal origin. There are 650–820 long, slender gill rakers, and head length is 24–29% of standard length (SL). Body is plain pale, greenish gray above and whitish below. This species reaches a maximum size of 105 cm and 50 kg (Luna, 1988).

Nutritional Facts

Proximate Composition (% WW)

Moisture[a]	Ash	Protein	Lipid
78.70	1.05	18.12	2.13

[a] g.

Source: Data from Ahmed et al. (2012).

Vitamin A and Minerals (per 100 g)

Vitamin A (RAE)	Ca[a]	Fe[b]	Zn[b]
<30		0.9	4.4	0

RAE, retinol activity equivalent.

[a] g.

[b] mg.

Source: Data from Thilsted (2012).

Major Class of Fatty Acids, EPA and DHA (% of Total Fatty Acids)

SFA	MUFA	PUFA	n-3	n-6	EPA	DHA
23.30	38.11	29.74	23.84	5.37	5.11	6.06

SFA: saturated fatty acid; MUFA: monounsaturated fatty acid; PUFA: polyunsaturated fatty acid; EPA: ecosapentaenoic acid; DHA: docosahexaenioic acid

Source: Ghomi et al. (2012).

Trace Elements (µg/g) in Fish Sample after Microwave Digestion

Al	Cr	Cu	Fe	Sr	Zn
6.50	0.26	0.65	12.2	0.52	12.6

Source: Kupeli et al. (2014).

Ctenopharyngodon idellus Valenciennes 1844

Order: Cypriniformes

Family: Cyprinidae

Common name: Grass carp, white amur.

139

Distribution: Bangladesh, China, Taiwan Province of China, Islamic Republic of Iran, the Laos, Myanmar, and Russian Federation (Weimin, 2014).

Habitat: Lakes, rivers, and reservoirs.

Description: Body of this species is elongated and cylindrical. Abdomen is rounded and compressed at the rear. Length of caudal peduncle is larger than the width. Head is medium with a terminal mouth that is arch shaped. Upper jaw extends slightly over lower jaw. Lateral line extends to caudal peduncle. Anus is close to anal fin. Body color is greenish yellow laterally and dorsal portion is dark brown. Abdomen is grayish white. This species reaches a maximum size of 1.4 m and 40 kg (Weimin, 2014).

Nutritional Facts

Proximate Composition (g/100 g Edible Portion, WW)

Energy[a]	Water	Protein	Fat	CHO[b]	Ash
104	79.2	17.6	3.7	0	1.0

[a] kcal.
[b] Carbohydrate.

Minerals (mg/100 g Edible Portion)

Ca	P	Fe	Na	K
21	179	1.1	49	178

Vitamins (per 100 g Edible Portion)

Retinol (µg)	Carotene (µg)	Thiamin (mg)	Riboflavin (mg)	Niacin (mg)	Vitamin C (mg)
39	0	0.01	0.07	1.6	2.2

Source: Data from Tee et al. (1989).

Hypophthalmichthys nobilis Richardson 1845 (= *Aristichthys nobilis*)

Order: Cypriniformes

Family: Cyprinidae

Common name: Bighead carp.

Distribution: Asia: China (Welcomme, 1988).

Habitat: Low flow areas of lakes, creeks, and channels.

Description: Body is laterally compressed and abdomen is rounded. Head is large and its length is larger than body height. Mouth is terminal and slanting upward. Lower jaw extends slightly over upper jaw. Lateral line extends to caudal peduncle. Tip of ventral fin reaches and exceeds anus. Body color is black in dorsal and upper lateral portion and silvery white in abdomen. Irregular black spots are seen on the lateral side of body. Fins are grayish in color. This species reaches a maximum size of 146 cm and 40 kg (Weimin, 2014).

Nutritional Facts

Proximate Composition (g/100 g Edible Portion, WW)

Energy[a]	Water	Protein	Fat	CHO[b]	Ash
142	74.4	16.2	8.6	0	1.1

[a] kcal.

[b] Carbohydrate.

Minerals (mg/100 g Edible Portion)

Ca	P	Fe	Na	K
25	192	0.9	50	211

Vitamins (per 100 g Edible Portion)

Retinol (µg)	Carotene (µg)	Thiamin (mg)	Riboflavin (mg)	Niacin (mg)	Vitamin C (mg)
12	0	0.01	0.08	2.4	4.9

Source: Data from Tee et al. (1989).

n-3 and n-6 Fatty Acids (%)

Σn-3	Σn-6	n-3/n-6
30.5	9.7	3.1

Source: Data from Steffens et al. (2006).

Tinca tinca Linnaeus 1758

Common name: Tench.

Distribution: Eurasia (Torres, 1991u).

Habitat: Freshwater, brackish water; shallow, densely vegetated lakes and backwaters.

Description: Body of this species is thickset, heavy, and laterally compressed. Caudal peduncle is characteristically deep and short. Skin is thickened and slimy. Scales are small and embedded. Head is triangular. Eye is orange-red and small. Snout is relatively long. The mouth is very oblique and has a pair of large maxillary barbels, one at each corner of the mouth. Overall coloration of body is olive-green and at times dark green or almost black, with golden reflections on ventral surface. This species grows to a maximum size of 70 cm and 7.5 kg (Torres, 1991u).

Nutritional Facts

Proximate Composition (% WW)

Moisture	Protein	Fat
80.23	18.29	0.58

141

Fatty Acids

MUFA (%)	PUFA (%)	n-3/n-6
27.90	46.92	1.93

PUFAs

Arachidonic (%)	Eicosapaentaenoic (%)	Docosohexaenoic (%)
8.92	7.97	13.87

Source: Data from Jankowska et al. (2006).

Major Elements in the Muscle (µg/100 g)

Na	K	Mg	Ca	P
33	369	51	58	207

Trace Elements in the Muscle (µg/100 g)

Fe	Mn	Se
843	95	0

Vitamins in the Muscle (g/100 g)

A	B1	B2	B6
1	75	180	0

Source: Data from Steffens et al. (2006).

Puntius chola Hamilton 1822

Order: Cypriniformes

Family: Cyprinidae

Common name: Swamp barb, chola barb.

Distribution: Asia: Pakistan, India, Nepal, Bangladesh, Sri Lanka, and Myanmar (Menon, 1999).

Habitat: Streams, rivers, canals, beels, baors, ponds, and inundated fields.

Description: Body of this species is deep and compressed. A single maxillary pair of barbels is present. Last simple dorsal ray is moderately strong and smooth. Lateral line is complete with 24–28 scales. Body is silvery with golden opercle. A dark blotch is present on the base of caudal fin. This species has a maximum size of only 15 cm (Vishwanath and Laisram, 2004).

Nutritional Facts

Proximate Composition (% WW)

Protein	Fat	Moisture	Ash
14.08	3.05	74.43	1.19

Source: Data from Mazumder et al. (2008).

Puntius gonionotus Bleeker 1850 (= *Puntius javanicus, Barbonemus gonionotus*)

Order: Cypriniformes

Family: Cyprinidae

Common name: Java barb, silver barb.

Distribution: Southeast Asia: Lower Xe Bangfai; Mekong Basin in Laos, Thailand, and Cambodia; Chao Phraya Basin; Malay Peninsula; Sumatra and Java (Rainboth, 1996).

Habitat: Midwater to bottom depths in rivers, streams, floodplains, and occasionally in reservoirs.

Description: Body is strongly compressed. Back is elevated and its dorsal profile is arched, often concave above the occiput. Head is small and snout is pointed. Mouth is terminal. Barbels are very minute or rudimentary, especially the upper ones, which sometimes disappear entirely. Color in life is silvery white, sometimes with a golden tint. Dorsal and caudal fins are gray to gray-yellow. Anal and pelvic fins are light orange and their tips are reddish. Pectoral fins are pale to light yellow. Very few tubercles are seen on the snout. This species grows to a maximum size of 90 cm and 13 kg (Bailly, 1997j; Rainboth, 1996).

Nutritional Facts

Proximate Composition (% WW)

Moisture	Ash	Protein	Lipid
79.23	1.03	15.74	3.99

Source: Data from Ahmed et al. (2012).

Puntius sophore F. Hamilton 1822 (= *Puntius stigma*)

Order: Cypriniformes

Family: Cyprinidae

Common name: Pool barb.

Distribution: Native to Asia; Pakistan, India, Nepal, Bangladesh, Myanmar, Bhutan, Afghanistan, and Yunnan, China.

143

Habitat: Freshwater and brackish water; rivers, streams, and ponds.

Description: Body of this species is moderately compressed. Dorsal profile is more convex than that of the abdomen. Mouth is small and terminal without barbels. Upper jaw is longer than lower jaw. Pectoral fin is as long as head excluding snout. Pelvic fin originates a little behind origin of dorsal. Lateral line is complete. A dark spot is seen at the tip of the tial and at the base of dorsal fin rays. Body when alive is silvery and back is grayish green to brownish. Flanks are with a bluish luster and abdomen is white. It reaches an adult size of 25 cm and a weight of 70 g (Zakia, 2010; Froese and Pauly, 2006a).

Nutritional Facts

Proximate Composition (% WW)

Moisture	Protein	Fat	Ash	CHO[a]
75.60	21.50	2.70	1.90	1.55

[a] Carbohydrate.

Source: Data from Musa (2009).

Vitamin A and Minerals (per 100 g)

Vitamin A (RAE)	Ca[a]	Fe[b]	Zn[b]
60	1.2	3.0	3.1

RAE, retinol activity equivalent.

[a] g.

[b] mg.

Source: Data from Thilsted (2012).

Puntius conchonius Hamilton 1822 (= *Barbus conchonius*)

Order: Cypriniformes

Family: Cyprinidae

Common name: Rosy barb, red barb.

Distribution: Subtropical parts of Asia: India, Bangladesh, Nepal, Pakistan, and Afghanistan; introduced into Singapore, Australia, Mexico, Puerto Rico, and Columbia (About home).

Habitat: Lakes and fast-flowing waters.

Description: In this species, rostral barbels are absent. Maxillary barbels are minute or absent. A stiff, serrated last unbranched dorsal fin ray and a black blotch on the caudal peduncle are present. Males have a brighter red coloration than the females, which look more gold or silver than red. Black blotches, spots, or bars are seen on the side of the body of both sexes. This species reaches a maximum size of only 15 cm (SerioulyFish, 2014f; About home).

144

Nutritional Facts

Proximate Composition (% DM)

Moisture	CHOª	Protein	Fat	Ash
4.54	21.52	13.55	28.12	1.00

ª Carbohydrate.

Amino Acids (% DM)

Methionine	Tryptophan
1.12	2.00

Source: Data from Ali et al. (2013).

Puntius ticto Hamilton 1822

Order: Cypriniformes

Family: Cyprinidae

Common name: Ticto barb, firefin barb, tic-tac-toe barb, two-spot barb.

Distribution: Pakistan, India, Nepal, Sri Lanka, Bangladesh (Dahanukar, 2010c.)

Habitat: Mostly montane and submontane regions, and floodplains.

Description: Mouth of this species is small and it is terminal in position. Barbels are absent. Depth of the body is less than one-third of the standard length. The ticto barb is silver and gold with two black spots, one just before the pectoral fin and one near the back tail. This species grows to a maximum size of only 10 cm (Bashar, 2011).

Nutritional Facts

Proximate Composition (% WW)

Moisture	Ash	Protein	Lipid
75.02	3.34	18.08	3.56

Source: Data from Ahmed et al. (2012).

Barbus barbus Linnaeus 1758

Order: Cypriniformes

Family: Cyprinidae

Common name: Common barbel.

Distribution: Native throughout Europe and China and has become established as an introduced species in Morocco and Italy.

Habitat: Bottom of rivers as well as open waters.

Description: Body of this species is slightly laterally compressed and lacks

145

an adipose fin. Lower lip is thick with a median swollen pad. Tip of dorsal is pointed and its posterior margin is concave. Pelvic originates below dorsal origin. It has a dark brown or gray mottled appearance and underside is light colored. Fins have a reddish tinge. This reaches a maximum size of 1.2 m and 12 kg (Torres, 1991a).

Nutritional Facts

Proximate Composition (g/100 g WW)

Moisture	Protein	Fat	Ash	CHO[a,b]
72.39	18.61	7.78	1.33	53.12

[a] mg/100 g.
[b] Carbohydrate.

Fatty Acids (%)

SFA	MUFA	PUFA	PUFA/SFA
28.63	45.27	26.31	0.92

Source: Data from Ljubojevic et al. (2013).

Barbus carnaticus Jerdon 1849 (= *Barbodes carnaticus, Puntius carnaticus*)

Order: Cypriniformes

Family: Cyprinidae

Common name: Carnatic carp.

Distribution: Endemic to the western Ghats of India (Manojkumar, 2006).

Habitat: Large pools in rivers and streams.

Description: Body of this species is elongated and its depth 2.5–3.4 times the standard length. Mouth is slightly subterminal and lips are moderately fleshy. Barbels are in two pairs. While maxillary pair is as long as the orbit, the rostral ones are much shorter. Dorsal fin is inserted slightly nearer to the tip of the snout than to the base of the caudal fin. Scales are fairly large. Lateral line is complete, with 28–32 scales. Color of body is olive green on back, fading to dull white glossed with gold on flanks and abdomen. Usually a faded band is seen above the lateral line. This species grows to a maximum size of 60 cm and 12 kg (Manojkumar, 2006).

Nutritional Facts

Proximate Composition (% WW)

Moisture	Fat	Protein	Ash
78.3	1.0	16.9	1.23

Source: Data from Natarajan and Srinivasan (1961).

Fatty Acids (%)

SFA	MUFA	PUFA
33.7	33.0	33.2

Source: Data from Gopakumar (1975).

Barbus filamentosus Valenciennes 1844
(= *Puntius filamentosus, Dawkinsia filamentosa*)

Order: Cypriniformes

Family: Cyprinidae

Common name: Black-spot barb, Indian tiger barb, mahecola, filament barb, longfin barb, featherfin barb.

Distribution: Asia: India, Sri Lanka, Burma, and possibly Thailand (Torres, 1991b).

Habitat: Rivers and tanks; hill streams and streams of the lowlands and wetlands.

Description: This species has an oblong and compressed body with equally convex dorsal and ventral profiles. Tips of dorsal rays are filamentous in matured males. A pair of rudimentary maxillary barbels is present. Body is olive dorsally with emerald green above the lateral line. Sides are silvery or yellow. Caudal fin is orange with white tips. Ventral fin is green. Other fins are orange. A large blotch is seen above the posterior part of the anal fin. This species attains a size of only 10 cm (Munro, 2000).

Nutritional Facts

Proximate Composition (% DM)

Protein	Fiber	Ash	Moisture	Energy[a]
33.5	0.4	10.0	8.3	363.7

[a] kcal/100 g.

Minerals (mg/g DM)

Fe	Cu	Ni	Na	K	Zn	Pb
0.074	0.034	0.028	3.0	6.8	0.07	0.018

Source: Data from Onyia et al. (2010).

Barbus grypus Heckel 1843

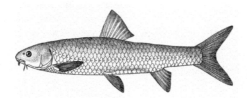

Order: Cypriniformes

Family: Cyprinidae

Common name: Shabbout.

Distribution: Asia: Tigris-Euphrates Basin (Binohlan, 1990a).

Habitat: Large to medium-sized rivers with moderate current.

147

Description: This species is cylindrical in appearance. It is dark olive, which looks lighter at the abdomen area. It has four whiskers and protruding plump lips. This species can grow to a length of nearly 2 m and a weight of over 50 kg (Crispina, 1990).

Nutritional Facts

Proximate Composition (g/100 g WW)

Protein	Fat	Moisture	Ash	CHO[a]	Energy[b]
19.34	4.04	74.65	1.06	0.88	117.5

[a] Carbohydrate.
[b] kcal/100 g.

Minerals (mg/100 g)

Cu	Zn	Fe	P	Ca
0.21	1.05	0.68	264.82	32.95

Fatty Acids (%)

SFA	MUFA	PUFA	PUFA/ SFA	PUFA n-3	PUFA n-6	n6/n-3
30.03	33.85	20.74	0.71	17.09	1.91	0.11

Amino Acids (mg/100 g)

Alanine	1015
Glycine	827.91
Valine	1018.03
Leucine	1473.02
Isoleucine	1007.62
Threonine	840.85
Serine	645.05
Proline	684.2
Aspartic acid	2462.3
Methionine	432.4
Hydrochloride proline	166.5
Glutamic acid	2385.2
Phenylalanine	663.8
Lysine	1425.1
Histidine	449.9
Tyrosine	727.3

Source: Data from Olgunoglu et al. (2011).

Schizopyge niger Heckel 1838

Order: Cypriniformes

Family: Cyprinidae

Common name: Chush snowtrout, Alghad snowtrout, ale snowtrout.

Distribution: Asia: India and Pakistan (Torres, 1991q).

Habitat: Lakes and adjoining channels.

Description: The species of this genus possesses a blunt snout and suctorial lip. It feeds on detritus and breeds in cold and clear pockets of lakes, particularly on the roots of willow trees. Eggs are either scattered in clutches along the bottom or adhered to submerged willow roots. This species grows to a maximum size of 27 cm and 2.7 kg (Torres, 1991q).

Nutritional Facts

Proximate Composition (% WW)

Moisture	Protein	Fat	Ash
71.58	16.43	7.42	2.28

Source: Data from Ganai (2012).

Schizothorax curvifrons Heckel 1838

Order: Cypriniformes

Family: Cyprinidae

Common name: Sattar snowtrout.

Distribution: Asia: Afghanistan, Pakistan, India, China; Iran, Uzbekistan, Kazakhstan, and Kyrgyzstan (Bailly, 1997k).

Habitat: Rivers, lakes, and swamps.

Description: The species of this genus possesses a blunt snout and suctorial lip. In this species, mature adults undertake spawning migration to incoming streams and breeding takes place amidst gravel and sandy beds. Spawning period is extremely protracted. Large specimens spawn earlier than the small ones. Diet consists of animals and plants. This species can reach a length of 56 cm and a weight of up to 1.3 kg (Bailly, 1997k).

Schizothorax esocinus Heckel 1838

Order: Cypriniformes

Family: Cyprinidae

Common name: Chirruh snowtrout.

Distribution: Asia: Afghanistan, Pakistan, India, Nepal, and China (Torres, 1991r).

Habitat: Mountain streams, rivers, and gravel-bottomed rivers.

Description: The species of this genus possesses a blunt snout and suctorial lip. It feeds on bottom detritus. Mature adults undertake spawning migration to incoming streams where they breed amidst gravel and sandy beds. Fry always occur in quiet parts of the streams or in the side branches of the main streams. This species reaches a maximum length of 47 cm (Torres, 1991r).

149

Schizothorax plagiostomus Heckel 1838

Order: Cypriniformes

Family: Cyprinidae

Common name: Khont.

Distribution: China, Afghanistan, Pakistan, Turkistan, Nepal, Ladkah, Tibet, Bhutan, and northeastern India (Froese and Pauly, 2006d).

Habitat: Rivers and tributaries.

Description: This species has an elongated subcylindrical body with a short, blunt, and slightly prognathous upper jaw. Ventral surface of head and anterior part of body are flattish, short, somewhat cone shaped, and blunt. Snout is usually smooth and covered with warys in the male. Interorbital space is broad and flat. Mouth is inferior, wide, and slightly arched. Lips are fleshy and continuous and marginally sharply attenuated. The margin of the lower lip is sharp. Barbels are in two pairs. Dorsal fin is inserted about opposite to pelvic fins. Caudal fin is deeply emarginated. Scales are very small and elliptical. This species reaches a maximum size of 60 cm and 2.5 kg (Froese and Pauly, 2006d).

Schizothorax labiatus McClelland 1842

Order: Cypriniformes

Family: Cyprinidae

Common name: Kunar snowtrout.

Distribution: Asia: India, Nepal, Pakistan, Afghanistan, and Tibet (China) (Froese and Pauly, 2006c).

Habitat: Hill streams.

Description: This species has a smooth and pointed snout. Mouth is subterminal, horizontal, arch shaped, and protractile. Lips are thick and fleshy. Lower labial fold is uninterrupted and trilobed. Median lobe is well developed and free at the tip. Barbels are in two pairs and are longer than the eye. Dorsal spine is strong and scales are small. Lateral line has 100–110 scales. Live specimens possess dark brown color with black specks on the back and yellowish white below. Fins are pinkish yellow. This species attains a maximum length of 30 cm (Froese and Pauly, 2006c; PMNH).

Nutritional Facts

Proximate Composition (% WW)

	Moisture	Protein	Fat	Ash
S. curvifrons	79.48	14.10	3.08	1.86
S. esocinus	77.84	15.88	2.09	2.60
S. plagiostomus	74.27	17.51	4.33	2.20
S. labiatus	76.52	17.22	1.88	1.97

Source: Data from Ganai (2012).

Schizothorax richardsonii Gray 1832

Order: Cypriniformes

Family: Cyprinidae

Common name: Common snowtrout, Alawan snowtrout.

Distribution: Asia: Himalayan region of India, Sikkim and Bhutan, Nepal, Pakistan, and Afghanistan (Luna, 1988s).

Habitat: Mountain streams and rivers.

Description: This species has a more slender body form and more length of caudal fin and snout (preorbital length) than percentage of total length. Mouth is semicircular and subventral in position. An adhesive organ occurs at the ventral head region, just below the mouth opening. It consists of a crescent callus part, and below it, labial folds bear semispherical tubercles. The values (against standard length) of different parameters are given in parentheses: snout length (5.46%) and caudal fin length (24.11%), eye diameter (4.71%), post-orbital length (7.02%), head length (17.05%), body depth (14.69%), dorsal fin base (9.74%), and anal fin base (5.17%). This species reaches a maximum size of 60 cm (Luna, 1988s; Dasi and Nag, 2008).

Nutritional Facts

Proximate Composition (% DM)

Moisture	Protein	Lipid	Ash
9.36	71.08	6.568	5.44

Macroelements (mg/100 g)

Ca	Mg	K	Na	P
10.05	97.50	280.50	38.50	369.59

Microelements (mg/100 g)

Cu	Co	Mn	Zn	Fe	Ni	Cr
0.291	1.416	0.666	2.229	4.2708	1.208	0.200

Source: Data from Hei and Sarojnalini (2012).

Neolissochilus hexagonolepis McClelland 1839

Common name: Katli, copper mahseer, chocolate mahseer.

Distribution: Nepal, Bangladesh, Pakistan, Myanmar, Thailand, Malaysia (Peninsular), China, Indonesia (Arunachalam, 2010).

Habitat: Streams with fast-flowing water mostly in high-gradient and low-gradient riffles and pools.

Description: Males of this species have greater head width at operculum, snout length, predorsal length, and preanal length than the females. On the other hand, the females have greater body depth at dorsal fin insertion, pectoral fin length, anal fin height, and ventral fin height than the males. Among all the dissimilar characteristics, the height of the anal fin is found significantly lower in males than in females. Snout length is the second most significantly different characteristic and is higher in males than in females. It grows to a maximum length of 120 cm and 11 kg (Laskar et al., 2013; Luna, 1988n).

Raiamas guttatus Day 1870

Common name: Burmese trout.

Distribution: Northeastern India (Nagaland), Myanmar, Thailand, Laos, Cambodia, Malay Peninsula, and China (Vishwanath, 2010).

Habitat: Shady areas and muddy bottoms in deep hill streams.

Description: This species has a single-rayed dorsal fin and cycloid scales. Both the jaws are toothless, and teeth are seen only in the throat. The distinguishing feature of this species is the presence of two rows of blue spots along its sides. It grows to a maximum length of only 30 cm, and its diet consists of insects and small fishes (ETYFish Project).

Tor putitora F. Hamilton 1822

Common name: Putitor mahseer, golden mahaseer.

Distribution: Himalayan region and elsewhere in South Asia and Southeast Asia, ranging from Afghanistan, Pakistan, India (Darjeeling to Kashmir), Nepal, Bangladesh, Bhutan, Sri Lanka, Myanmar, western Iran to eastern Thailand (Jha and Rayamajhi, 2010b).

Habitat: Montane and submontane regions, in streams and rivers.

Description: Body of this species is elongated. Both dorsal and ventral profiles are straight and somewhat compressed. Mouth is small and upper jaw is slightly longer than lower jaw. Lips are thick and fleshy. There are two pairs of barbels. Pelvic fins contain a scaly appendage. Caudal fin is deeply forked. Side of the body is greenish silvery. Belly is silvery to white. Scales are golden with dark base and formed of minute black spots. Anal, pelvic, and pectoral fins are reddish yellow. This species grows to a maximum size of 2.75 m and 54 kg (Nayan, 2011f).

Tor tor Hamilton 1822

Common name: Mahseer.

Distribution: Myanmar, Bhutan, Bangladesh and Pakistan, and India (Rayamajhi et al., 2010).

Habitat: Rivers with a rocky bottom.

Description: Head of this species is slightly shorter than depth. Dorsal profile is more sharply arched than ventral profile. Lips are thick and fleshy with a continuous labial fold across the lower jaw. Mouth is small and its gape does not extend below the eyes. Snout is pointed and jaws are of about the same length. Two pairs of barbels are seen and maxillary ones are slightly longer than rostral ones but shorter than the eye. Dorsal fin is opposite to or slightly in advance of ventral fin. Pectoral is reaching pelvic. Caudal fin is deeply forked and lateral line is complete. This species grows to a maximum size of 152 cm and 78 kg (Desai, 2003).

153

Nutritional Facts

Proximate Composition (% DM)

	Moisture[a]	Protein	Lipid	Ash
N. hexagonolepis	29.90	78.20	13.70	5.30
R. guttatus	13.20	75.90	14.90	5.20
T. putitora	14.70	77.17	13.90	6.20
T. tor	14.80	74.75	15.90	7.00

[a] Fresh weight.

Source: Data from Kosygin et al. (2001).

Tor tambroides Bleeker 1854

Common name: Malayan mahseer, Thai mahseer.

Distribution: Throughout southern Asia and from Indonesia to China (Entri, 2013).

Habitat: Rivers, lakes, and reservoirs.

Description: Adults of this species are streamlined in shape, and vary in color from yellowish brown to pinkish, with dark fins. The tail fin is deeply forked and the dorsal fin is large. Scale color is reddish white. Lip is thick and fleshy. The presence of a long median lobe on the lower lip and large scales with dark vertical bands, especially on the scales above the lateral line, is the most distinctive feature of this species. The ranges of gill rakers and vertebrae numbers are 17–23 and 30–34, respectively. The number of branchiostegals rays is three. The meristic characteristics showed greater differences among the *Tor tambroides* populations than the morphometric characteristics. This fish grows to a maximum length of 100 cm and maximum weight of 80 kg (Poh, 2006; Entri, 2013).

Nutritional Facts

Proximate Composition (% DM)

Moisture[a]	Protein	Ash	Fat	Fiber	NFE[b]
74.45		48.57	10.02	32.40	0.70	3.83

[a] Wet weight.

[b] Nitrogen-free extract.

Source: Data from Ismail et al. (2013).

Fatty Acids of Muscle (% of Total Fatty Acids)

14:0	3.35
16:0	25.97
16:1	4.83
18:0	10.63
18:1n-9	23.40
18:3n-3	0.18
20:0	2.38
20:3	0.73
20:4n-6	5.94
24:1	0.10
20:5n-3	3.33
22:5n-3	1.50
22:6n-3	6.73
SFA	42.32
MUFA	31.35
PUFA n-3	11.51
n-3/n-6	0.79

Source: Data from Ramezani-Fard et al. (2011).

Tor khudree Sykes 1839

Order: Cypriniformes

Family: Cyprinidae

Common name: Black mahseer, khudree mahseer, Deccan mahseer, yellow mahseer.

Distribution: India and Sri Lanka (Raghavan, 2013).

Habitat: Major rivers and reservoirs; cool, fast-flowing, rocky streams; tanks and cold-water lakes.

Description: In this species males are generally smaller than females. Males have large labial palp. There is no black streak behind gill opening. Tip of snout is not fleshly tuberculated. Lateral line has 25–27 scales. Length of head is more or less equal to depth of body. Eye diameter is 5.5–7.0 in head length. Lips are thick, with an uninterrupted fold across the lower jaw. Maxillary pair of barbels are longer than the rostral, and extend to below the last third of the eye. Pectoral fin is as long as the head excluding the snout; it reaches the ventral fin, which is a little shorter. Caudal fin is deeply forked. Color of body is silvery or greenish along the upper half of the body, becoming silvery shot with gold on the sides and beneath. Lower fins are reddish yellow. This species grows to a maximum size of 130 cm and 45 kg (Big Fishes of the World).

Nutritional Facts

Proximate Composition (per 100 g DM)

Protein (g)	Lipid (mg)	Moisture[a] (%)
19.40	03.80	76.50

[a] Wet weight.

Source: Data from Patole (2012).

Rasbora daniconius Hamilton 1822

Order: Cypriniformes

Family: Cyprinidae

Common name: Slender barb, blackline rasbora, common rasbora, slender rasbora, striped rasbora.

155

Distribution: Mekong, Chao Phraya, and Salween Basins, the northern Malay Peninsula, and westward to the Indus and Sri Lanka; Indonesia (Borneo and Sumatra) (Jenkins and Ali, 2013).

Habitat: Freshwater and brackish water; ditches, ponds, canals, streams, rivers, and inundated fields.

Description: Body of this species is oblong and compressed. The greatest width of the head equals its post-orbital length. Lateral line is complete with only the last few scales lacking pores. Overall it has a silver body with olive-colored undertones, but the scales are large and highly reflective, giving a sparkling effect. The belly of the male is yellow or reddish, while the female's is whiter. However, its most prominent feature is a blue-black line that runs the body length. This line is finely out-lined in a bright gold. It reaches 15 cm in length (Froese and Pauly, 2006b; Animal-world).

Nutritional Facts

Proximate Composition (per 100 g DM)

Protein (g)	Lipid (mg)	Moisture[a] (%)
17.90	01.47	81. 20

[a] Wet weight.
Source: Data from Patole (2012).

Rasbora tornieri Ahl 1922

Order: Cypriniformes

Family: Cyprinidae

Common name: Yellow tail rasbora.

Distribution: Asia: Indochina, Malaysia, and Indonesia (Luna, 1988q).

Habitat: Streams, canals, and ditches in lowland floodplains.

Description: This species has a broad, dark brown, sharply defined midlateral body stripe extending from the opercle to the caudal fin base. Body depth is 4.2–4.6 times in standard length. Lateral line is complete. There are one or two scale rows between the lateral line and midventral scale rows in front of the pelvic fin. A carmine red caudal fin with or without a narrow dark margin is seen. This species reaches a maximum size of only 17 cm (Luna, 1988q; SeriouslyFish, 2014g).

Nutritional Facts

Vitamin A and Minerals (per 100 g)

Vitamin A (RAE)	Ca[a]	Fe[b]	Zn[b]
374	0.7	0.7	2.7

RAE, retinol activity equivalent.
[a] g.
[b] mg.
Source: Data from Thilsted (2012).

Esomus danricus Hamilton 1822

Order: Cypriniformes

Family: Cyprinidae

Common name: Flying barb.

Distribution: Asia: Pakistan, India, Nepal, Bangladesh, Afghanistan, Sri Lanka, and Myanmar (Luna, 1988h).

Habitat: Freshwater and brackish water; ponds, weedy ditches, and irrigation canals.

Description: This species is known for its extremely long barbels, which reach almost to the anal fin. It is a silver fish with a black line on an elongated body and gold fins. This species reaches a maximum length of 15 cm.

Nutritional Facts

Proximate Composition (% WW)

Moisture	Ash	Protein	Lipid
77.20	2.54	17.12	3.13

Source: Data from Ahmed et al. (2012).

Vitamin A and Minerals (per 100 g)

Vitamin A (RAE)	Ca[a]	Fe[b]	Zn[b]
890	0.9	12.0	4.0

RAE, retinol activity equivalent.

[a] g.

[b] mg.

Source: Data from Thilsted (2012).

Esomus longimanus Lunel 1881

Order: Cypriniformes

Family: Cyprinidae

Common name: Mekong flying barb.

Distirbution: Asia: Mekong from the Khorat Plateau, Thailand, to the Great Lake, Cambodia (Casal, 1995b).

Habitat: Ditches, canals, and ponds often in areas with extensive growth of submerged aquatic plants; stagnant water bodies, including sluggish flowing canals as well as medium to large rivers.

Description: This species has two pairs of barbels, and the rostral barbel extends well beyond the eye. The first pectoral ray sometimes reaches the anal fin. A narrow dark line runs from the head to the caudal fin base. This species reaches a maximum size of only 8 cm (Casal, 1995b).

157

Nutritional Facts

Vitamin A and Minerals (per 100 g)

Vitamin A (RAE)	Ca[a]	Fe[b]	Zn[b]
415	0.8	11.3	4.9

RAE, retinol activity equivalent.

[a] g.

[b] mg.

Source: Data from Thilsted (2012).

Neolissochilus stracheyi Day 1871

Order: Cypriniformes

Family: Cyprinidae

Common name: Stracheyi mahseer, blue mahseer, waterfall carp.

Distribution: Asia: Myanmar through Thailand, Cambodia, and Vietnam (Luna, 1980o).

Habitat: Clear forested streams and rivers; swift-flowing streams.

Description: This species has nine branched dorsal fin rays and a black lateral stripe. The last simple dorsal ray is smooth and nonosseous. On the side of the snout and below the eye, there is a large patch of tubercles. Post-labial groove is found interrupted medially. Color when alive is bronze back and silvery belly. This species has a maximum length of 60 cm (Luna, 1980o).

Nutritional Facts

Proximate Composition (% DM)

Moisture	Protein	Lipid	Ash
15.77	57.373	7.512	4.43

Macroelements (mg/100 g)

Ca	Mg	K	Na	P
9.35	68.75	236.25	60.87	106.48

Microelements (mg/100 g)

Cu	Co	Mn	Zn	Fe	Ni	Cr
2.565	4.125	1.000	0.4375	8.000	1.375	0.200

Source: Data from Hei and Sarojnalini (2012).

Semiplotus manipurensis Vishwanath & Kosygin 2000

Order: Cypriniformes

Family: Cyprinidae

Common name: Assamese kingfish.

Distribution: Asia: Manipur, India (Garilao, 1995c).

Habitat: Flowing hill streams with rocky beds; shallow and fast-flowing waters; deeper, slower waters.

Description: This species has a broad body (width 17.3–22.1% SL). The last dorsal spine is not serrated. Dorsal and pelvic fin rays are branched. There are 12–13 predorsal scales. Dorsal fin base length is 34.0–39.7% SL. There are seven scale rows between the dorsal fin origin and lateral line. Many horny tubercles are found scattered on each side of the snout tip, which extends posteriorly to the region below the anterior margin of the orbit. This species has a maximum length of 18.5 cm (Garilao, 1995c).

Nutritional Facts

Proximate Composition (% DM)

Moisture	Protein	Lipid	Ash
13.40	69.00	7.348	5.43

Macroelements (mg/100 g)

Ca	Mg	K	Na	P
10.75	81.00	284.24	65.87	973.15

Microelements (mg/100 g)

Cu	Co	Mn	Zn	Fe	Ni	Cr
0.875	2.187	ND	2.062	8.375	1.500	ND

ND, no data.

Source: Data from Hei and Sarojnalini (2012).

Abramis brama Linnaeus 1758

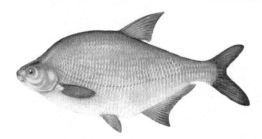

Order: Cypriniformes

Family: Cyprinidae

Common name: Freshwater bream.

Distribution: Europe and Asia (Luna, 1988a).

Habitat: Warm and shallow lakes, rivers, backwaters, and brackish estuaries.

159

Description: It is a deep-bodied fish with a high back and flattened sides. Head is comparatively small. Mouth is subinferior and it can be extended as a tube. Eyes are small. Color of body is dark brown or grayish on the back. Fins are gray or light brown, and those underneath are reddish tinted. It grows to a maximum length of 75 cm and weight of 9.1 kg (FAO, 2014b; Luna, 1988a).

Nutritional Facts

Proximate Composition (g/100 g WW)

Moisture	Protein	Fat	Ash	CHO[a,b]
78.66	17.59	3.24	0.80	41.93

[a] mg/100 g.
[b] Carbohydrate.

Fatty Acids (g/100 g)

SFA	MUFA	PUFA	PUFA/SFA
27.27	56.09	17.07	0.63

Source: Data from Ljubojevic et al. (2013).

Amblypharyngodon mola Hamilton 1822

Order: Cypriniformes

Family: Cyprinidae

Common name: Mola carplet.

Distribution: Asia: Pakistan, India, Bangladesh, Myanmar, Afghanistan (Talwar and Jhingran, 1991).

Habitat: Ponds, canals, beels, slow-moving streams, and paddy fields.

Description: Body of this species is laterally compressed. Dorsal profile is more convex than that of the ventral. Barbels are absent. Caudal fin is deeply forked and its lobes are pointed. Dark markings are present in dorsal and anal fins. Body color is light greenish on the back and silvery sides and beneath. It grows to a maximum length of 20.0 cm (Galib, 2010a).

Nutritional Facts

Proximate Composition (% WW)

Moisture	Fat	Protein	Ash
74.72	5.83	3.56	1.4

Minerals (µg/g DM)

Fe	Cu	Zn	Ca	Ni	Mn	K	Mg	Na
4.82	0.05	1.09	8.95	0.08	0.35	1.93	2.26	0.74

Source: Data from Devi and Sarojnalini (2012).

Vitamin A and Minerals (per 100 g)

Vitamin A (RAE)	Ca[a]	Fe[b]	Zn[b]
2680	0.9	5.7	3.2

RAE, retinol activity equivalent.
[a] g.
[b] mg.
Source: Data from Thilsted (2012).

Leptobarbus hoevenii Bleeker 1851

Order: Cypriniformes

Family: Cyprinidae

Common name: Hoven's carp, Sultan fish, mad barb.

Distribution: Thailand, Laos, Cambodia, Vietnam; Peninsular Malaysia and the Greater Sunda Islands of Sumatra and Borneo (SeriouslyFish, 2014d).

Habitat: Large rivers, lakes, streams, and floodplains.

Description: Body of this species is sleek and built for speed. Caudal fin is deeply forked. It is a plainly colored fish with a dark blotch behind the gill cover. Young fish (<10 cm SL) exhibit a faint lateral stripe with distinctive orange-red finnage, a more rounded head shape, different eye position, shorter barbels, and more rounded caudal fin lobe. It has a maximum length of 1 m (SeriouslyFish, 2014d).

Hampala macrolepidota Kuhl & van Hasselt 1823

Order: Cypriniformes

Family: Cyprinidae

Common name: Hampala barb.

Distribution: Brunei, Malaysia (Peninsular, Sarawak, and Sabah), and Indonesia (Kalimantan, Sumatra, and Java) (Bailly, 1997e).

Habitat: Clear rivers or streams with running water and sandy to muddy bottoms; found in most water bodies except small creeks, torrents, and shallow swamps.

Description: This species has a black bar between the dorsal and pelvic fins in adults. Its orange to red caudal fin has a black longitudinal, marginal stripe along each lobe. Juveniles usually have an additional vertical bar on the caudal peduncle. They also possess a black teardrop-shaped marking on the cheek. Barbel is always longer than eye width. Eyes are located in the upper side of the head. It reaches a maximum length of 70 cm (Bailly, 1997e).

161

Osteochilus vittatus Valenciennes 1842 (= *Osteochilus hasseltii*)

Order: Cypriniformes

Family: Cyprinidae

Common name: Hard-lipped barb, bonylip barb.

Distribution: Southeast Asia, including major rivers in Burma, Thailand, Indochina, and Peninsular Malaysia. It is also recorded from Sumatra, Java, and Borneo. In Singapore (Ecology Asia).

Habitat: Large streams with slow current and muddy to sandy substrate.

Description: Body of this species is laterally compressed and rhomboidal shaped. It is mainly silvery in color, but its dorsal surface is sometimes greenish or brownish. Fins are pinkish grading to red at the margins. Extreme margins of pelvic and anal fins are pale. Caudal fin is deeply forked, and barbels are small. At the base of the caudal fin, there is a single black marking. It has a maximum length of 32 cm (Ecology Asia).

Nutritional Facts

Proximate Composition (% WW)

	Moisture	Protein	Fat	Ash	CHO[a]
L. hoevenii	80.0	16.4	1.2	1.0	1.4
H. macrolepidota	77.0	17.5	2.4	1.0	1.2
O. vittatus	80.0	16.5	0.9	1.2	1.4

[a] Carbohydrate.

Amino Acids (g/16 g N)

	L. hoevenii	*H. macrolepidota*	*O. vittatus*
Lysine	16.4	7.0	7.5
Histidine	11.8	2.9	1.0
Arginine	4.0	5.4	9.3
Aspartic acid	8.5	7.5	8.0
Threonine	9.6	3.5	5.5
Serine	4.3	3.2	3.9
Glutamic acid	10.7	10.7	18.2
Proline	3.2	4.1	2.4
Glycine	5.7	5.5	4.8
Alanine	5.6	5.9	5.9
Cystine	0.5	0.9	0.5

Continued

	L. hoevenii	*H. macrolepidota*	*O. vittatus*
Valine	6.8	4.1	5.4
Methionine	4.4	2.6	2.7
Leucine	3.1	5.4	7.7
Isoleucine	6.2	3.1	6.7
Tyrosine	3.7	2.8	3.6
Phenylalanine	2.6	3.4	4.1

Source: Data from Zanariah and Rehan (1988).

Aspius aspius Linnaeus 1758

Order: Cypriniformes

Family: Cyprinidae

Common name: Asp.

Distribution: Europe and Asia (Bailly, 1997a).

Habitat: Lakes and lower reaches of rivers and estuaries; large rivers.

Description: Body of this species is oblong and laterally compressed with a long sharp head. It has an anal fin with 12–14.5 branched rays. Lateral line has a total of 64–76 scales. Maxilla extends beyond the front margin of the eye, and a sharp keel is seen between the pelvic fin and anal fin origins covered by scales. Coloration is green back with silver to blue tints. Flanks are lighter and belly is silver white. Pectoral, pelvic, and anal fins are gray to brown. This species has a maximum length of 120 cm and weight of 9 kg (Bailly, 1997a).

Nutritional Facts

Proximate Composition (g/100 g WW)

Moisture	Protein	Fat	Ash	CHO[a, b]
78.51	18.07	2.78	1.16	36.26

[a] mg/100 g.
[b] Carbohydrate.

Fatty Acids (%)

SFA	MUFA	PUFA	PUFA/SFA
27.99	47.69	24.60	0.88

Source: Data from Ljubojevic et al. (2013).

Leptocypris niloticus Joannis 1835 (= *Barilius niloticus*)

Order: Cypriniformes

Family: Cyprinidae

163

Common name: Nile minnow.

Distribution: From Senegal to Ethiopia, and north as far as the Nile delta (Geelhand, 2013c).

Habitat: Running waters, especially on sandy shores and in irrigation canals.

Description: This species has a body depth of 3.8–5.7 times and head length 3.7–4.5 times the SL (standard length). Eye diameter is 2.8–3.6 times the head length. There are 12 scales around the caudal peduncle, 9–12 anal fin branched rays, and 8–9 dorsal fin branched rays. Sides are plain or with irregular markings and back is dark. This species reaches a maximum size of 9.5 cm and only 10 g (Geelhand, 2013c).

Nutritional Facts

Proximate Composition (% DM)

Protein	Fiber	Ash	Moisture	Energy[a]
37.1	0.42	5.0	8.0	370.12

[a] kcal/100 g.

Minerals (mg/g)

Fe	Cu	Ni	Na	K	Zn	Pb
0.086	0.042	0.02	3.7	8.4	0.082	0.017

Source: Data from Onyia et al. (2010).

Barilius bendelisis Hamilton 1807

Order: Cypriniformes

Family: Cyprinidae

Common name: Hamilton's barila.

Distribution: Asia: Pakistan, India, Nepal, Bangladesh and Sri Lanka, Bhutan, and Myanmar (Luna, 1988c).

Habitat: Streams and rivers along the base of hills with pebbly and rocky bottom.

Description: Body of this species is elongated and compressed. A moderately cleft mouth has two pairs of short barbes (maxillary and rostral). Rostral pair may be either rudimentary or absent. Dorsal profile is less convex than ventral. A thick layer of spiny tubercles is seen on snout and lower jaw. Dorsal fin is inserted in advance of anal fin and is nearer to the caudal base than to the snout tip. It grows to a maximum length of 22.7 cm (Das et al., 2012; Fahad, 2012).

Nutritional Facts

Proximate Composition (% WW)

Ash	Protein	Fat
2.50	14.90	3.56

Source: Data from Jha et al. (2014).

Systomus sarana Hamilton 1822 (= *Puntius sarana*)

Order: Cypriniformes

Family: Cyprinidae

Common name: Olive barb.

Distribution: Native to Afghanistan, Pakistan, India, Nepal, Bangladesh, Bhutan and Sri Lanka; Thailand and Myanmar.

Habitat: Sandy bed mixed with mud and in fairly swift current.

Description: Body of this species is deep and moderately compressed. Dorsal profile is elevated. Eyes are large and situated in the anterior half of the head, and snout is rounded. Two pairs of nostrils are present, and the nostrils of each pair are separated by a muscular flap.

Mouth is wide, and pores are absent on the snout. Two pairs of barbels are present. Rostral barbels are slightly shorter than the maxillary pair. Pelvics originate below the origin of the dorsal fin. Color of body is silvery in the back. Opercle is shot with gold and yellowish and white in the abdomen. It reaches a maximum size of 42 cm (Rahman, 1989).

Nutritional Facts

Proximate Composition (% WW)

Moisture	Protein	Fat	Ash
71.39	16.73	9.00	2.02

Source: Data from Begum and Minar (2012).

Rutilus frisii Nordmann 1840

Order: Cypriniformes

Family: Cyprinidae

Common name: Kutum.

Distribution: Eurasia: Black and Azov Sea Basins (Reyes, 2011).

Habitat: Freshwater and brackish water; large estuaries and their large, freshened plume waters; coastal lakes connected to rivers and lowland stretches of large rivers.

Description: Body of this species is almost cylindrical with a depth of 19–26% standard length. Abdomen is posterior to pelvic, which is rounded. Snout is rounded and stout. Mouth is subterminal. Dorsal fin has 9–10.5

branched rays. Iris and fins are gray or slightly yellowish. Breeding males possess large, scattered tubercles on top and side of head. It reaches a maximum size of 70 cm and 8 kg (Reyes, 2011).

Nutritional Facts

Proximate Composition (g/100 g WW)

Moisture	Protein	Fat	Ash
73.51	15.16	4.36	1.77

Source: Data from Ghomi et al. (2011).

Fatty Acids (% Total Fatty Acids)

			n-3	n-6		
SFA	MUFA	PUFA	PUFA	PUFA	EPA	DHA
18.99	47.67	20.25	18.17	1.50	3.95	8.85

EPA, eicosapentaenoic acid; DHA, docosahexaenoic acid.

Source: Data from Ghomi et al. (2012).

Devario aequipinnatus McClelland 1839 (= *Danio aequipinnatus*)

Order: Cypriniformes

Family: Cyprinidae

Common name: Giant danio.

Distribution: India, Nepal, Bangladesh, Myanmar, and Thailand (Sharpe, 2014).

Habitat: Hill streams and standing waters.

Description: It is a deep-bodied species. Body is iridescent gold with steel blue-colored spots and stripes running lengthwise from the gills to the tail. In females, stripe bends upwards at the base of the tail, while in males, this stripe runs straight, extending through the tail. Fins are pale golden in color and rounded, while the tail fin is forked. There are several color variations, including an albino one. This species reaches a size of only 10 cm (Sharpe, 2014).

Nutritional Facts

Proximate Composition (per 100 g DM)

Protein (g)	Lipid (mg)	Moisture[a] (%)
15.00	02.70	77.00

[a] Wet weight.

Source: Data from Patole (2012).

Garra mullya Sykes 1839

Order: Cypriniformes

Family: Cyprinidae

Common name: Sucker fish, stone sucker.

Distribution: Asia: Nepal (Reyes, 1993d).

Habitat: Streams and rivers.

Description: This species has very well-developed cycloid scales over its body, but its thorax on the ventral surface is scaleless and smooth. Pectoral and pelvic fins are more ventrally attached to body. Fin rays are segmented and branched distally. Anterior lip is densely beset with tubercles. It has an adhesion apparatus consisting of a sucking disc as part of its mentum (the protruding part of the lower jaw), along with modified lips and paired fins, which enable the fish to adhere to rocky substrates in torrential streams. This species has a maximum size of 17 cm (Saxena, 1959).

Nutritional Facts

Proximate Composition (per 100 g DM)

Protein (g)	Lipid (mg)	Moisture[a] (%)
17.85	3.60	75.90

[a] Wet weight.

Source: Data from Patole (2012).

Osteobrama cotio cotio **Hamilton 1822**

Order: Cypriniformes

Family: Cyprinidae

Common name: Cotio.

Distribution: Asia: Pakistan, India, Nepal, and Bangladesh (Luna, 1988p).

Habitat: Rivers, lakes, ponds, and ditches.

Description: Body of this species is deeply compressed. Dorsal and ventral profiles are equally convex. Mouth is small and terminal without barbels. Upper jaw is longer than the lower jaw. Gill opening is wide. Body is silvery with dark on back. Fins are light greenish in color. Lateral line is present and complete. This species reaches a maximum size of 15 cm (Galib, 2010c).

Nutritional Facts

Proximate Composition (per 100 g DM)

Protein (g)	Lipid (mg)	Moisture[a] (%)
16.50	1.40	82.30

[a] Wet weight.

Source: Data from Patole (2012).

Salmostoma phulo **Hamilton 1822 (= *Salmophasia phulo*)**

Order: Cypriniformes

Family: Cyprinidae

Common name: Finescale razorbelly minnow, silver fish.

Distribution: Asia: India and Bangladesh (Torres, 1991m; Bashar, 2013).

Habitat: Rivers, canals, beels.

Description: This species has an elongated body that is laterally compressed. Eyes are situated at the anterior part of head. Abdominal profile cuts behind the base of the pectoral. Lower jaw extends up to the front margin of the orbit. Dorsal fin begins opposite the origin of the anal fin. Pectoral fin does not reach pelvic fin. Caudal fin is deeply forked. Lateral line is complete and curved gently downward. Body is silvery with a bright and silvery lateral band. It grows to a maximum length of 12 cm (Bashar, 2013).

Proximate Composition (% WW)

Moisture	Ash	Protein	Lipid
75.86	3.25	17.23	3.66

Source: Data from Ahmed et al. (2012).

Salmophasia bacaila Hamilton 1822 (= *Oxygaster bacaila*)

Order: Cypriniformes

Family: Cyprinidae

Common name: Large razorbelly minnow.

Distribution: Pakistan, northern India, Bangladesh, Nepal, and Afghanistan (Devi and Dahanukar, 2013).

Habitat: Freshwater and brackish water; slow-running streams, rivers, ponds, and inundated fields.

Description: Body of this species is elongated and strongly compressed. Mouth is oblique. Lower jaw has a well-developed symphysial knob. Scales are very small. There is a good space between anal and caudal fins. Lateral line is concave. Body is darkish dorsally and rest of the body is silvery. It grows to a maximum size of 17.5 cm (Bashar, 2010b).

Nutritional Facts

Proximate Composition (% DM)

Moisture	CHO[a]	Protein	Fat	Ash
3.08	20.22	13.75	18.43	0.99

[a] Carbohydrate.

Amino Acids (% DM)

Methionine	Tryptophan
0.99	0.98

Source: Ali et al. (2013).

Parachela siamensis Günther 1868 (= *Oxygaster oxygastroides*)

Order: Cypriniformes

Family: Cyprinidae

Common name: Glass fish, glass barb.

Distribution: Asia: Mekong and Chao Phraya Basins and Indonesia (Ortañez, 2005b).

Habitat: Lowland large rivers, including streams, lakes, and flooded fores.

Description: Body of this species is oblong and strongly compressed. Profile of back is nearly straight and concave behind elevated snout. Abdomen is convex with a sharp keel. Lateral line is descending gradually, in a slight curve. Body is yellowish brown with a silvery hue. A silvery longitudinal band is seen along sides. Fins are more or less powdered with black, forming a distinct longitudinal band on each caudal lobe. It reaches a maximum size of 20 cm (Weber and deBeaufort, 1962).

Nutritional Facts

Vitamin A and Minerals (per 100 g raw cleaned parts)

Vitamin A (RAE)	Ca[a]	Fe[b]	Zn[b]
480	0.6	1.2	2.2

RAE, retinol activity equivalent.
[a] g.
[b] mg.
Source: Data from Thilsted (2012).

Garra gatyla J.E. Gray 1830

Order: Cypriniformes

Family: Cyprinidae

Common name: Sucker head.

Distribution: Widespread in South Asia; Pakistan, India (Rayamajhi and Jha, 2010b).

Habitat: Fast-flowing streams with boulders and rocks along the Himalayan ranges.

Description: Body of this species is elongated and subcylindrical to compressed. Mouth is arched to semicircular or inferior. A suctorial disc is seen on the chin. Snout is with well-developed median proboscis. Two pairs of barbels are present. Dorsal fin is inserted nearer to tip of snout. It reaches a length of 14.5 cm (Fahad, 2011c).

Garra annandalei Hora 1921

Order: Cypriniformes

Family: Cyprinidae

Common name: Annandale garra, Tunga garra.

Distribution: India (northern Bengal, Bihar, and Assam), eastern and central Nepal (Kosi drainage), and Bangladesh (Rayamajhi and Jha, 2010a).

Habitat: Swift and clear mountain streams with rocks and bounders.

Description: Body of this species is elongated and cylindrical with a slightly depressed head. Mouth is small and semicircular. Proboscis is absent and two pairs of barbels are seen. Pectoral is shorter than head and caudal fin is deeply emarginated. Body color is grayish and dorsal side is dark gray. Belly is pale. There is a black spot on the upper angle of the gill opening. Lower lobe of caudal fin is black and other fins are blackish (Fahad, 2011b).

Garra lissorhynchus McClelland 1842

Order: Cypriniformes

Family: Cyprinidae

Common name: Khasi garra.

Distribution: Asia: Northeastern India (Singh, 2010).

Habitat: Hills stream with rocky bed.

Description: It is a species of *Garra* with no proboscis on snout. There are 32–34 lateral line scales and 11–14 predorsal scales. Breast and belly are naked. Distance of vent from anal fin is 2.5–4.6 times the interdistance between the pelvic fin origin and anal fin. A black spot is seen at the upper angle of the gill opening and a broad black W-shaped band on the anterior half of the caudal fin. It reaches a maximum length of 7.4 cm (India Biodiversity Portal).

Nutritional Facts of Muscle (mg/g WW)

	Protein	Free Amino Acids	Phosphorus
Garra gatyla	168.80	0.975	30.00
Garra annandalei	166.30	0.930	29.66
Garra lissorhynchus	164.89	0.908	33.60

Source: Data from Bhagowati and Ratha (1982).

FAMILY: NEMACHELIDAE (RIVER LOACHES)

Acanthocobitis botia Hamilton 1822

Order: Cypriniformes

Family: Nemacheilidae

Common name: Mottled loach.

Distirbution: Asia: Pakistan and China (Reyes, 1993a).

Habitat: Clear water; swift-flowing streams with rocky, pebbly, and sandy bottoms.

Description: This species is without a suborbital flap in the male, and this flap is replaced by a suborbital groove. Lateral line reaches at least to the anus. It is a beautiful, slender, and elongated fish with a decorative pattern, especially as a juvenile. It has a grayish green background color with a lighter, whitish belly. Dark blotches are seen on the sides and top. Dorsal and tail fins have stripes made up of dark spots. Under great care in an aquarium, it can also have an orange-red cast to its fins. It reaches a maximum size of only 12 cm (Reyes, 1993a; Animal-world).

Nutritional Facts

Proximate Composition (per 100 g DM)

Protein (g)	Lipid (mg)	Moisture[a] (%)
22.70	01.70	79. 85

[a] Wet weight.
Source: Data from Patole (2012).

FAMILY: COBITIDAE (LOACHES)

Lepidocephalus guntea Hamilton 1822

Order: Cypriniformes

Family: Cobitidae

Common name: Guntea loach.

Distribution: Northern parts of South Asia (Allen, 2012).

Habitat: Flowing and clear standing waters; rivers, beels, and streams.

Description: This species has a rounded/truncate caudal fin. A scaleless patch is seen on top of head. Flanks have spotted patterning in females and a solid, dark lateral stripe in males. Caudal fin possesses dark, reticulated markings. A black spot is present at the base of the caudal fin. This species reaches a maximum length of 10.4 cm (Galib, 2013).

171

Nutritional Facts

Proximate Composition (% WW)

Protein	Lipid	Moisture	CHOᵃ	Ash
17.60	17.40	75.43	0.12	9.20

ᵃ Carbohydrate.

Amino Acids Liberated from Fish Protein in Enzyme Actions

Amino Acid	Pepsin	Pepsin + Trypsin
Leucine	+	+
Phenylalanine	+	+
Valine	+	+
Methionine	+	+
Butyric acid	+	+
Alanine	–	–
Threonine	+	+
Tyrosine	+	+
Glutamic acid	–	–
Glycine	–	–
Arginine	+	+
Serine	–	–
Aspertic acid	–	–
Histidine	–	+
Lysine	+	+
Cystine	+	+

Source: Data from Sarojnalini (2010).

Lepidocephalichthys thermalis Valenciennes 1846 (= *Lepidocephalus thermalis*)

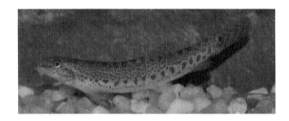

Order: Cypriniformes

Family: Cobitidae

Common name: Common spiny loach.

Distribution: Asia: India and Sri Lanka (Reyes, 1993q).

Habitat: Quiet, flowing waters with a sandy substrate.

Description: This species has a truncate/rounded caudal fin. There are no scales on top of the head. Dark, squarish spots are seen on the flanks. A total of three to five broad dark bars are visible on the caudal fin. Dorsal fin origin is anterior to pelvic fin origin. Barbels are usually relatively small. This species reaches a maximum length of 38 cm (SeriouslyFish, 2014c).

Nutritional Facts

Proximate Composition (per 100 g DM)

Protein (g)	Lipid (mg)	Moistureᵃ (%)
15.25	02.30	81.95

ᵃ Wet weight.

Source: Data from Patole (2012).

FAMILY: BOTIIDAE (DWARF LOACH, DWARF CHAIN LOACH)

Syncrossus berdmorei Blyth 1860 (= *Botia berdmorei*)

Order: Cypriniformes

Family: Botiidae

Common name: Redfin tiger loach.

Distribution: Thailand, Burma, and India.

Habitat: Streams and rivers.

Description: In this species, there are laterally orientated rows of large dark markings that run across the entire body and head. In young individuals, these markings appear first in the region behind the opercle and extend further along the body as they mature. There is also a strong red coloring in the fins and tail. This species reaches a size of 25 cm (SeriouslyFish, 2014h).

Nutritional Facts

Proximate Composition (% WW)

Protein	Lipid	Moisture	CHO[a]	Ash
18.21	7.05	71.50	0.12	8.12

[a] Carbohydrate.

Amino Acids Liberated from Fish Protein in Enzyme Actions

Amino acids	Pepsin	Pepsin + Trypsin
Leucine	–	–
Phenylalanine	+	+
Valine	+	+
Methionine	+	+
Butyric acid	+	+
Alanine	–	–
Threonine	+	+
Tyrosine	+	+
Glutamic acid	–	–
Glycine	–	–
Arginine	+	+
Serine	+	+
Aspertic acid	–	–
Histidine	–	+
Lysine	+	+
Cystine	–	–

Source: Data from Sarojnalini (2010).

ORDER: PERCIFORMES (PERCHES AND ALLIES)

FAMILY: ANABANTIDAE (CLIMBING GOURAMIES OR CLIMBING PERCHES)

Anabas testudineus Bloch 1792

Order: Perciformes

Family: Anabantidae

Common name: Climbing perch.

Distribution: Asia: India to Wallace line, including China (Froese and Pauly, 2012).

Habitat: Rivers and pond waters.

Description: Body of this species is covered by cycloid scales. Lateral line sense organ is identified by the black spots, and there is a conspicuous one at the base of the caudal fin. Color when alive is dark to pale greenish, very pale below, and back dusky to olive. Head has longitudinal stripes ventrally. Posterior margin of opercle has a dark spot. Iris is golden reddish in color. It grows up to 25 cm and is a very hardy fish, due to the presence of an accessory respiratory organ. Rosette-like structures are found very close to the pectoral fin (Froese and Pauly, 2012).

Nutritional Facts

Proximate Composition (g/100 g WW)

	Moisture	Protein	Fat	Ash	CHO[a]	Energy[b]
Native fish	70.26	18.05	8.64	1.30	1.74	156.92
Hybrid fish	65.82	20.22	11.28	0.98	1.59	189.16

[a] Carbohydrate.
[b] kcal.

Minerals (mg/100 g WW)

	Native Fish	Hybrid Fish
Zn	1.07	0.57
Na	125.44	85.13
Ca	367.80	310.16
K	161.26	129.35
Fe	2.10	1.91
Cu	0.47	0.32
Mg	26.32	18.84
Al	5.95	3.29

Nonessential Amino Acids (g/100 g WW)

	Native Fish	Hybrid Fish
Aspartic acid	0.58	0.64
Serine	0.74	0.86
Glutamic acid	0.39	0.42
Glycine	2.16	2.70
Alanine	0.84	0.82
Tyrosine	0.19	0.26
Arginine	1.48	1.68

Source: Data from Monalisa et al. (2013).

Essential Amino Acids (g/100 g WW)

	Native Fish	Hybrid Fish
Lysine	2.57	2.40
Leucine	0.08	0.11
Isoleucine	0.52	0.58
Valine	0.12	0.11
Methionine	0.17	0.21
Threonine	0.22	0.27
Histidine	0.99	1.12

FAMILY: SCATOPHAGIDAE (SCATS)

Scatophagus argus argus Linnaeus 1766

Order: Perciformes

Family: Scatophagidae

Common name: Spotted scat

Distribution: Indo-Pacific: Kuwait to Fiji, north to southern Japan, south to New Caledonia (Valdestamon, 2013).

Habitat: Marine, freshwater, brackish water; harbors, natural embayments, brackish estuaries, and the lower reaches of freshwater streams, mangroves.

Description: Body of this species is quadrangular and strongly compressed. Dorsal head profile is steep. Eye is moderately large and its diameter somewhat smaller than the snout length. Snout is rounded. Mouth is small, horizontal, and not protractile. Teeth are villiform, arranged in several rows on jaws. This species occurs in two basic color morphs called green scat and ruby or red scat. Juveniles have a few large roundish blotches, about the size of the eye or about five or six broad, dark, vertical bars. In large adults, spots may be faint and restricted to the dorsal part of flanks. This species grows to a maximum length of 38 cm (Valdestamon, 2013).

Nutritional Facts

Proximate Composition and Fatty Acids (g/100 g WW)

Moisture	Protein	Fat	Fatty Acid	Ash	CHO[a]	Energy[b]
80.73	12.99	2.13	1.49	1.09	3.06	83.37

[a] Carbohydrate.
[b] kcal/100 g.

Minerals (µg/100 g)

Cu	Zn	Fe	Mn
1838.38	3108.31	1064.45	82.05

Source: Data from Islam et al. (2010).

FAMILY: SCIAENIDAE (DRUMS, CROAKERS)

Aplodinotus grunniens Rafinesque 1819

Order: Perciformes

Family: Sciaenidae

Common name: Freshwater drum.

Distribution: North and Central America (Ortañez, 2005a).

Habitat: Medium to large rivers and lakes.

Description: This species is deep-bodied and equipped with a long dorsal fin divided into two sections. Dorsal fin has 10 spines and 29–32 rays. Body is silvery in color without distinctive tail fin spot. It grows to a maximum size of 95 cm and 24.7 kg (Texas Parks and Wildlife Foundation, 2014).

Nutritional Facts (per 3 oz WW):

Calories	101.2 kcal
Total fat	4.2 g
Saturated fat	1 g
Polyunsaturated fat	1 g
Monounsaturated fat	1.9 g

Cholesterol	54.4 mg
Sodium	63.8 mg
Potassium	233.8 mg
Carbohydrate	0 g
Dietary fiber	0 g
Protein	14.9 g
Vitamin A	144.5 IU
Vitamin C	0.9 mg
Thiamin	0.1 mg
Riboflavin	0.1 mg
Niacin	2 mg
Vitamin B6	0.3 mg
Folate	12.8 µg
Folic acid	0 µg
Vitamin B12	1.7 µg
Pantothenic acid	0.6 mg
Calcium	51 mg
Iron	0.8 mg
Magnesium	25.5 mg
Zinc	0.6 mg
Copper	0.2 mg
Phosphorus	153 mg
Selenium	10.7 µg

Source: Data from Mealographer (2006).

Pseudotolithus typus Bleeker 1863

Order: Perciformes

Family: Sciaenidae

Common name: Longneck croaker.

Distribution: Eastern Atlantic: Mauritania to Angola, becoming scarce north of Cape Verde (Maigret and Ly, 1986).

Habitat: Coastal waters from shoreline to about 150 m depth, over mud and sandy mud bottoms, uncommon in rocky areas; juveniles and subadults enter estuaries and rivers.

Description: Body of this species is very long and rounded in cross section. Head is long and clearly concave on nape. Interorbital space is very narrow, less than eye. Mouth is large, strongly oblique, and its lower jaw is clearly projecting. A pair of sharp canines is seen near the tip of the upper jaw. Caudal fin is pointed. Color of the body is silvery gray, often with faint dark lines following oblique scale rows on the back and upper sides. Pectoral, anal, and pelvic fins are yellowish and caudal is dark gray. This species reaches a size of 140 cm and 15 kg (Marine Species Identification Portal).

Nutritional Facts

Proximate Composition (% WW)

Protein	Lipid	Moisture	Ash
19.82	1.06	75.39	1.16

Major Amino Acids (mg/g)

Glutamic acid	31.58
Aspartic acid	31.58
Lysine	18.20
Leucine	16.45
Arginine	11.32
Alanine	11.32

Medium Amino Acids (mg/g)

Valine	8.98
Isoleucine	8.59
Glycine	7.61
Threonine	8.43
Serine	8.21
Phenylalanine	7.06
Proline	6.18
Methionine sulfone	5.80

Minor Amino Acids (mg/g)

Histidine	3.91
Cystine	2.12
Tyrosine	1.18
Ornithine	0.52
Taurine	1.65
g-Aminobutyric acid	0.45
Hydroxyproline	ND

ND, not detected.

Source: Data from Osibona (2011).

Pseudolithus elongatus Bowdich 1825

Order: Perciformes

Family: Sciaenidae

Common name: Bobo croaker.

Distribution: Eastern Atlantic: Senegal to southern Angola (Garilao, 1995b).

Habitat: Marine and brackish water; coastal waters over mud bottom from shoreline to about 50 m depth; also enters estuaries.

Description: Body of this species is elongated and compressed. Scales on body are mostly ctenoid (comb-like), and scales on breast and head are smaller and cycloid. Lateral line extends to tip of caudal fin. Head is short and eye is rather large. Mouth is large, strongly oblique. Of the two jaws either the terminal or lower jaw is slightly projecting. Teeth are small and set in narrow bands in jaws. Chin is without barbel but with six pores. Among these pores, the medial pair is located at the tip of the lower jaw. Snout has five marginal pores only. The gill rakers are long and slender. Caudal fin is pointed (juveniles) to rhomboidal (adults). Color of body is silvery gray with a reddish tint, often with oblique lines and scattered dark spots on the back. Belly is yellowish during breeding season. Fins are grayish and dark spots are seen on the soft part of the dorsal fin, forming two or three longitudinal rows. Tip of first dorsal fin is dusky, and pelvic and anal fins are yellowish. It grows to a maximum size of 47 cm (FAO, 2014f).

Nutritional Facts

Proximate Composition (% WW)

Moisture	Protein	Fiber	CHO[a]	Ash	Energy[b]
72.51	22.85	0.80	2.49	2.38	0.60

[a] Carbohydrate.
[b] J/kg DM.
Source: Data from Anene et al. (2013).

FAMILY: CHANNIDAE (SNAKEHEADS)

Channa marulius Hamilton 1822 (= *Ophiocephalus marulius*)

Order: Perciformes

Family: Channidae

Common name: Great snakehead, bullseye snakehead.

Distribution: Asia: India to China, south to Thailand and Cambodia (Reyes, 1993b).

Habitat: Sluggish or standing water in canals, lakes, and swamps.

Description: Body of this species is elongated and almost rounded or cylindrical in cross section. Mouth is large and eyes are moderate. Pectoral fin is half of head length and pelvic fin is about 75% of pectoral length. Caudal fin is rounded. Body color is gray or grayish green on lateral line. There are four or five large black blotches in adults. Body is also distinct with white spots scattered on body and fins. This species grows to a maximum size of 183 cm and 30 kg (Fahad, 2011a).

Nutritional Facts

Proximate Composition (% WW)

Moisture	Ash	Protein	Lipid
81.42	0.6	16.19	1.79

Source: Data from Ahmed et al. (2012).

178

Major Amino Acids (% of Total Protein)

Glycine	35.77
Alanine	10.19
Lysine	9.44
Aspartic acid	8.53
Proline	6.86

Major Fatty Acids (% of Total Fatty Acids)

Oleic acid (C18:1)	22.96
Stearic acid (C18:0)	15.31
Linoleic acid (C18:2)	11.45
Arachidonic acid (C20:4)	7.44

Source: Data from Zakaria et al. (2007).

Minerals (mg/kg)

Ca	Fe	Mg	Zn
16.5	5.26	216	10.32

Source: Data from Paul et al. (2013).

Vitamin A and Minerals (per 100 g Raw Cleaned Parts)

Vitamin A (RAE)	Ca[a]	Fe[b]	Zn[b]
200	1.4	1.5	1.5

RAE, retinol activity equivalent.

[a] g.

[b] mg.

Source: Data from Thilsted (2012).

Channa micropeltes Cuvier 1831

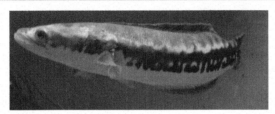

Order: Perciformes

Family: Channidae

Common name: Indonesian snakehead, giant snakehead, giant mudfish.

Distribution: Asia: Mekong and Chao Phraya Basins; the Malay Peninsula, and the islands of Sumatra and Borneo (Reyes, 1993c).

Habitat: Lowland river and swamp.

Description: It has an eel-like body and its dorsal fin is long, stretching across the back. Head is small with a big mouth displaying rows of sharp teeth for trapping prey. A broad, dark longitudinal stripe in seen in adults, and two black longitudinal stripes with a bright orange intermediate area in juveniles.

This species grows to a maximum size of 50 cm and 20 kg. This species has a physiological need to breathe atmospheric air, forcing it to the surface to trap air in its suprabranchial organ (Reyes, 1993c).

Nutritional Facts

Proximate Composition (g/100 g Edible Portion, WW)

Energy[a]	Water	Protein	Fat	CHO[b]	Ash
31	78.8	19.7	0.2	0.1	1.2

[a] kcal.

[b] Carbohydrate.

Minerals (mg/100 g Edible Portion)

Ca	P	Fe	Na	K
32	214	0.4	16	335

Vitamins (per 100 g Edible Portion)

Retinol	Carotene	Thiamin	Riboflavin	Niacin	Vitamin C
19	0	0.02	0.09	2.2	1.2

Source: Data from Tee (1989).

Minerals (per 100 g Raw Cleaned Parts)

Ca[a]	Fe[b]	Zn[b]
1.1	1.2	1.4

RAE, retinol activity equivalent.

[a] g.

[b] mg.

Source: Data from Thilsted (2012).

Channa punctatus Bloch 1793

Order: Perciformes

Family: Channidae

Common name: Spotted snakehead.

Distribution: Afghanistan, eastward through Khyber Pass into Indus River Basin, Pakistan; rivers of the plains of India; Sri Lanka; southern Nepal; Bangladesh; Myanmar; eastward to Yunnan Province, southwestern China (USGS, 2012a).

Habitat: Stagnant waters and muddy streams on plains; muddy waters up to 600 m.

Description: Body of this species is elongated and mostly rounded. Pelvic fins are more than half the length of pectoral fins, extending to anal fin. Pectoral fins are plain without vertical bands. Caudal fin is rounded. Mouth is large and lower jaw has three to six canines behind a single row of villiform teeth. Colors of live fish vary from black to pale green on the dorsum, and sides and ventral sides are white to pale yellow, sometimes with a red tinge. Several dark blotches are seen on lower sides, and occasionally black spots are present on the body and dorsal, anal, and caudal fins. Dorsal, anal, and caudal fins are dark gray, sometimes with a reddish edge. Pectoral and pelvic fins are pale orange. This species grows to a maximum size of 31 cm (USGS, 2012a).

Nutritional Facts (mg/g WW)

CHO[a]	Glycogen	Protein	Free Amino Acids	Lipid	Free FA
9.4	2.7	74.4	6.3	38.9	14.3

FA, fatty acid.

[a] Carbohydrate.

Source: Data from Janakiram et al. (1983).

Proximate Composition (% WW)

Moisture	Protein	Lipid	Ash	CHOᵃ
34.43	41.38	3.21	20.74	0.24

ᵃ Carbohydrate.

Source: Data from Islam et al. (2013).

**Vitamin A and Minerals
(per 100 g Raw Cleaned Parts)**

Vitamin A (RAE)	Caᵃ	Feᵇ	Znᵇ
140	0.8	1.8	1.5

RAE, retinol activity equivalent.

ᵃ g

ᵇ mg.

Source: Data from Thilsted (2012).

Channa striata Bloch 1793 (= *Ophiocephalus striatus*)

Order: Perciformes

Family: Channidae

Common name: Striped snakehead.

Distribution: From Sri Lanka to Indonesia; Philippines, China, and Cambodia (FAO, 2014c).

Habitat: Survives dry season by burrowing into bottom mud of lakes, canals, and swamps.

Description: Body is elongated. Head is broad and flattened. Top and sides of head are covered with scales. Mouth is large. Only small teeth are seen on palate. Eyes are in anterior part of head. Lateral line has 42–57 scales. Dorsal fin is longer than anal fin. Caudal fin is rounded. Oblique bars are seen on body. This species grows to a maximum size of 100 cm and 7 kg (FAO, 2014c).

Nutritional Facts

Proximate Composition (% WW)

Moisture	Protein	Fat	Ash	CHOᵃ
79.6	16.1	1.1	1.1	2.1

ᵃ Carbohydrate.

Amino Acids (g/16 g N)

Lysine	16.4
Histidine	2.4
Arginine	5.1
Aspartic acid	7.1
Threonine	3.6
Serine	3.4
Glutamic acid	12.8
Proline	2.8
Glycine	3.4
Alanine	4.4
Cystine	0.8
Valine	3.1
Methionine	2.9
Leucine	5.8
Isoleucine	3.8
Tyrosine	2.9
Phenylalanine	3.2

Source: Data from Zanariah and Rehan (1988).

Minerals (mg/100 g)

K	Na	Ca	P	Fe
242.1	63.7	108.3	456.5	0.5

Source: Data from Wimalasena and Jayasuriya (1996).

Parachanna obscura Günther 1861 (= *Channa obscura*)

Order: Perciformes

Family: Channidae

Common name: African obscure snakehead.

Distribution: Africa: Nile, Senegal to the Chad system, up to the Congo system (Sanciangco, 2002).

Habitat: Marginal vegetation and floodplains.

Description: Head of this species is depressed anteriorly, relatively long, and covered with large scales. Lower jaw is slightly longer than upper jaw, with four to six well-developed canines. Eyes are laterally located and large and help the fish to locate its prey quickly. Lateral line is typically complete, rarely discontinous. Coloration is distinct in having a series of dark blotches, some of which may coalesce, and no chevron-shaped bars are seen across the middle of the back. This species grows to a maximum size of 50 cm and only 1 kg (USGS, 2012b).

Nutritional Facts

Proximate Composition (% WW)

Moisture	Protein	Ash	Fat	CHO[a]
68.61	18.23	2.68	0.19	6.93

[a] Carbohydrate.

Source: Data from Oluwafunmilola and Ogunkoya (2006).

FAMILY: AMBASSIDAE (ASIATIC GLASS FISHES)

Chanda nama Hamilton 1822

Order: Perciformes

Family: Ambassidae

Common name: Elongate glass perchlet.

Distribution: India, Nepal, Pakistan, and Bangladesh (Dahanukar, 2010a).

Habitat: Freshwater and brackish water; standing and running waters.

Description: Body of this species is well compressed and laterally flat. Dorsal and ventral profiles are equally convex. Lower jaw is longer than the upper jaw. Lateral line is partly distinct and partly absent. Second dorsal spine is distinct. Scales are minute and rounded. Caudal fin is forked. Body is transparent and yellowish white with many small black dots. First dorsal and tip of second dorsal fins are deep black. Caudal fin is black and orange. This species grows to a maximum size of only 11 cm (Sultana, 2010a).

Nutritional Facts

Proximate Composition (% WW)

Protein	Fat	Moisture	Ash
18.26	1.53	65.88	3.92

Source: Data from Mazumder et al. (2008).

Pseudambassis baculis Hamilton 1822

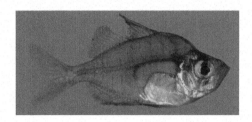

Order: Perciformes

Family: Ambassidae

Common name: Himalayan glassy perchlet.

Distribution: India (Himalayan and Indo-Gangetic Plains), Nepal and Bangladesh (Dahanukar, 2010b).

Habitat: Ponds, ditches, pools, and rivers.

Description: This species has a small, slender body with an oblique mouth. Scales are small. Lower jaw is almost equal to upper jaw. Color of the body is translucent with a yellowish green back. Flank and belly are silvery white. Fins are hyaline. A golden spot is seen on the occiput. It reaches a maximum size of only 5 cm (Galib, 2011).

Nutritional Facts

Proximate Composition (% WW)

Moisture	Ash	Protein	Lipid
78.62	2.92	15.60	2.86

Source: Data from Ahmed et al. (2012).

Parambassis ranga Hamilton 1822

Order: Perciformes

Family: Ambassidae

Common name: Indian glassy fish.

Distribution: Asia: Pakistan, India, Bangladesh, Myanmar, Thailand, Malaysia, and Nepal (Reyes, 1993j).

183

Habitat: Freshwater and brackish water; sluggish and standing waters and impoundments.

Description: It is a relatively deep-bodied and laterally compressed species. While the head and belly are silvery, the rest of the body is transparent, so the prominent backbone and other bones are visible. In this species, fins are long and rounded with the exception of two separate, pointed dorsal fins. The caudal fin is moderately long and forked. It has a pale green iridescence, particularly over the dorsal area. It grows to a maximum size of 8 cm.

Nutritional Facts

Vitamin A and Minerals (per 100 g Raw Cleaned Parts)

Vitamin A (RAE)	Ca[a]	Fe[b]	Zn[b]
1679	1.0	1.8	2.3

RAE, retinol activity equivalent.
[a] g.
[b] mg.
Source: Data from Thilsted (2012).

Parambassis wolffii Bleeker 1850

Order: Perciformes

Family: Ambassidae

Common name: Duskyfin glassy perchlet.

Distribution: Asia: Discontinuous distribution in mainland and insular Southeast Asia: Chao Phraya and Mekong Basins; islands of Sumatra and Borneo; Maeklong, Peninsular, and Southeast Thailand (Bailly, 1997i).

Habitat: Main rivers and lakes; sluggish rivers and floodplains.

Description: The short, deep body is characteristic of the species. The second anal fin spine is enlarged, and there are 43–46 scales in the lateral line series. This species grows to a maximum size of 20 cm (Bailly, 1997i).

Vitamin A and Minerals (per 100 g Raw Cleaned Parts)

Vitamin A (RAE)	Ca[a]	Fe[b]	Zn[b]
260	1.1	1.4	1.6

RAE, retinol activity equivalent.
[a] g.
[b] mg.
Source: Data from Thilsted (2012).

FAMILY: MORONIDAE (TEMPERATE BASSES)

Morone saxatilis Walbaum 1792

Order: Perciformes

Family: Moronidae

Common name: Striped bass.

Distribution: Western Atlantic (Kareen, 2005).

Habitat: Freshwater, marine water; diadromous.

Description: It is a highly prized sportfish, which, although anadromous (meaning it migrates between salt and freshwater), can live its full life in freshwater. It is a sleek silver fish that sports dark longitudinal stripes and is also known as Atlantic striped bass, striper, linesider, rock, pimpfish, or rockfish. Striped bass are one of six species belonging to the Moronidae family (temperate basses); they are not related to the black basses (which are in the sunfish family Centrarchidae). Striped bass commonly reach a length of 120 cm (3.9 ft) and are thought to live up to 30 years (EOL).

Nutritional Facts

Proximate Composition (per 100 g Edible Portion, WW)

Water (g)	Energy (kcal)	Protein (g)	Fat (g)
79.22	97	17.73	2.33

Amino Acids (g/100 g)

Tryptophan	0.199
Threonine	0.777
Isoleucine	0.817
Leucine	1.441
Lysine	1.628
Methionine	0.525
Cystine	0.190
Phenylalanine	0.692
Tyrosine	0.599
Valine	0.914
Arginine	1.061
Histidine	0.522
Alanine	1.072
Aspartic acid	1.816
Glutamic acid	2.647
Glycine	0.851
Proline	0.627
Serine	0.723
Total amino acids	17.10

Lipids (per 100 g)

SAF (g)	MUFA (g)	PUFA (g)	EPA (mg)	DHA (mg)	Cholesterol (mg)
0.507	0.660	0.784	169	585	80

Minerals (mg/100 g)

Ca	Fe	Mg	P	K	Na	Zn	Cu	Mn	Se[a]
15.00	0.84	40	198	256	69	0.40	0.03	0.02	36.5

[a] µg.

185

Vitamins (per 100 g)

Vitamin C (mg)	0.0
Thiamin (mg)	0.10
Riboflavin (mg)	0.03
Niacin (mg)	2.10
Pantothenic acid (mg)	0.75
Vitamin B6 (mg)	0.30
Folate (µg)	9

Choline (mg)	ND
Vitamin B12 (µg)	3.82
Vitamin A (IU)	90
Vitamin E (mg)	ND
Vitamin D(IU)	ND
Vitamin K (µg)	ND

ND, no data.

Source: Data from Tacon and Metian (2013).

Dicentrarchus labrax **Linnaeus 1758**

Order: Perciformes

Family: Moronidae

Common name: European seabass.

Distribution: Eastern Atlantic: Norway to Morocco, the Canary Islands and Senegal; Mediterranean and Black Sea (Binohlan, 1990b).

Habitat: Estuaries and brackish water lagoons. Sometimes they venture upstream into freshwater.

Description: Body of this species is rather elongated. Opercle has two flat spines. Mouth is terminal and moderately protractile. Two separate dorsal fins are seen. The first has 8–10 spines and the second has 1 spine. Anal fin has three spines. Scales are small. Lateral line is complete but not extending onto caudal fin. Caudal fin is moderately forked. Color of the body is silvery gray to bluish on the back and silvery on the sides, the belly is sometimes tinged with yellow. Young ones possess some dark spots on the upper part of the body. This species grows to a maximum size of 103 cm and 12 kg (FAO, 2014d).

Nutritional Facts

Proximate Composition (% WW)

Moisture	Ash	Protein	Lipid
77.38	1.17	19.75	1.22

Source: Data from Beklevk et al. (2005).

Amino Acids (g/100 g)

Arginine	8.36
Histidine	2.43
Isoleucine	4.14
Leucine	7.21
Lysine	7.61
Methionine	2.58
Cysteine	1.00
Phenylalanine	4.46
Tyrosine	3.90
Threonine	4.29
Tryptophan	0
Valine	4.55

Source: Data from Jarmolowicz and Zakês (2014).

FAMILY: GOBIIDAE (GOBIES)

Glossogobius giuris Hamilton 1822

Order: Perciformes

Family: Gobiidae

Common name: Bareye goby, freshwater goby, bar-eyed goby, flat-headed goby, flathead goby.

Distribution: Africa to Oceania: Red Sea and East Africa and most inland freshwater bodies over the Indian Ocean and western Pacific; from austral Africa and Madagascar to India and south of China (Luna, 1988j).

Habitat: Clear to turbid freshwater to estuarine habitats in rivers and streams with sand, gravel, or rock substrate.

Description: Head of this species is flattened and lower jaw is projecting. Body is pale without longitudinal lines. Dorsal fins have small spots forming longitudinal stripes. Pelvic fins are jointed but attached to the body only from their anterior part. Body is brownish yellow with five or six dark rounded spots on its sides. Individuals living on dark substrates may be very dark. Some living on very light substrates show an ivory coloration. Dorsal fins are light with brownish spots. Pelvic fins are gray. Pectoral and caudal fins are gray and often hyaline. This species grows to a maximum size of 5 cm (Luna, 1988j).

Nutritional Facts

Proximate Composition (g/100 g WW)

Moisture	CHO[a]	Lipid	Protein	Ash
81.3	2.6	0.1	12.8	1.0

[a] Carbohydrate.

Minerals (mg/100 g WW)

K	Na	Ca	P	Fe
316.1	77.0	336.6	485.5	1.3

Source: Data from Wimalasena and Jayasuriya (1996)

Amino Acids (% DM)

Methionine	Tryptophan
1.26	0.90

Source: Data from Ali et al. (2013).

FAMILY: LATIDAE (LATES PERCHES)

Lates niloticus Linnaeus 1758

187

Order: Perciformes

Family: Latidae

Common name: Nile perch.

Distribution: Ethiopian region of Africa; Lake Mariout, near Alexandria. Exists in Lakes Albert, Victoria, Rudolph, Nyanza, and Tana. Oubanguhi River at Bangui, French Congo; central North America (Froese and Pauly, 2003b).

Habitat: Freshwater channels, lakes, and irrigation canals; brackish waters of lakes.

Description: Body of this species is elongated, compressed, and deep. Head has concave snout profiles. Opercle has one spine. Lateral line extends onto caudal peduncle, reaching the posterior margin of the fin. Pelvic axis has scaly process. Dorsal fin is bipartite with seven or eight spines on the first part. One spine and 8–11 soft rays are on the second part. Anal fin has three spines and six to nine soft rays. Caudal fin is rounded. It grows to a maximum size of 2 m and 200 kg (Froese and Pauly, 2003b).

Nutritional Facts

Proximate Composition (g/100 g DM)

Lipid	Protein	Moisture[a]
6.8	77.9	75.6

[a] Wet weight.

Minerals (mg/g DM)

K	P	Ca	Na	Mg	Zn[a]	Fe[a]	Cu[a]
11,550	7270	2305	2395	705	62	26	0.9

[a] µg.

Essential Amino Acids (EAAs) (µg/g DM)

Leucine	244.1
Isoleucine	151.1
Valine	167.3
Threonine	153.9
Lysine	281.1
Histidine	105.3
Methionine	106.9
Phenylalanine	139.1
Total EAAs	1348.8

Source: Data from Mohamed et al. (2010).

FAMILY: CICHLIDAE (CICHLIDS)

Oreochromis aureus Steindachner 1864

Order: Perciformes

Family: Cichlidae

Common name: Blue tilapia.

Distribution: Africa and Eurasia (Casal, 1995c).

Habitat: Warm ponds and impoundments as well as lakes and streams.

Description: This species has a narrow preorbital bone (depth maximum of 21.5% of head length in fishes up to 21.3 cm standard length). There is no enlargement of the jaws in mature fish (lower jaw not exceeding and usually less than 36.8% head length). Caudal is without regular dark vertical stripes but with a broad pink to bright red distal margin. Breeding males assume an intense bright metallic blue on the head, a vermilion edge to the dorsal fin, and a more intense pink on the caudal margin. This species grows to a maximum size of 45.7 cm and 2 kg (Casal, 1995c).

Nutritional Facts

Proximate Composition (%)

Dry Matter	Protein[a]	Lipid[a]	Ash[a]
24.50	20.20	3.01	1.22

[a] Wet weight basis.

Amino Acids (g/100 g)

Threonine	3.49
Valine	4.23
Methionine	2.40
Isoleucine	3.90
Leucine	6.16
Phenylalanine	3.45
Histidine	2.20
Lysine	7.80
Arginine	4.71
Total EAAs	**38.36**
Serine	4.31
Glutamic acid	6.86
Proline	3.63
Glycine	3.62
Alanine	5.27
Aspartic acid	7.39
Tyrosine	1.90
Total NEAAs	**32.97**

EAA, essential amino acid; NEAA, nonessential amino acid.

Source: Data from Taşbozan et al. (2013b).

Oreochromis mossambicus Peters 1852

Order: Perciformes

Family: Cichlidae

Common name: Mozambique tilapia.

Distribution: Tropical and subtropical habitats around the globe; native of East Africa to Natal; Illovo, Mazoe, and Zambezi Rivers; Mozambique, Rhodesia, Natal; Japan, Taiwan, Southeast Asia, and India.

Habitat: Warm, weedy pools of sluggish stream, canals, and ponds.

Description: Body of this species is deep and laterally compressed with long dorsal fins, the front part of which have spines. Scales are cycloid. A knob-like protuberance is present behind the upper jaw on the dorsal surface of the snout. Upper jaw length shows sexual dimorphism, and the mouth of males

189

is larger than that of females. Lateral line is interrupted. Spinous and soft ray parts of the dorsal fin are continuous. Caudal fin is truncated. Native coloration is a dull greenish or yellowish, and there may be weak banding. During spawning season, pectoral, dorsal, and caudal fins become reddish. Adults reach a size of 35 cm and 1.13 kg.

Nutritional Facts

Proximate Composition (g/100 g Edible Portion, WW)

Energy[a]	Water	Protein	Fat	CHO[b]	Ash
96	78.9	16.4	3.0	0.8	0.9

[a] kcal.
[b] Carbohydrate.

Minerals (mg/100 g Edible Portion)

Ca	P	Fe	Na	K
22	147	0.6	48	253

Vitamins (µg per 100 g Edible Portion)

Retinol[a]	Carotene[a]	Thiamin	Riboflavin	Niacin	Vitamin C
21 µg	0 µg	0 mg	0.30 mg	1.6 mg	0.6 mg

[a] µg.

Source: Data from Tee et al. (1989).

Oreochromis niloticus Linnaeus 1758 (= *Tilapia niloticus*)

Order: Perciformes

Family: Cichlidae

Common name: Nile tilapia.

Distribution: Native to Africa from Egypt south to East and Central Africa, and as far west as Gambia. It is also native to Israel, and numerous introduced populations exist outside its natural range (e.g., Brazil) (Froese and Pauly, 2005).

Habitat: A wide variety of freshwater habitats.

Description: Body of this species is compressed. Caudal peduncle's depth is equal to length. Scales are cycloid. A knob-like protuberance is absent on the dorsal surface of the snout. Upper jaw length shows no sexual dimorphism. Lateral line is interrupted. Spinous and soft ray parts of dorsal fin are continuous. Caudal fin is truncated. During spawning season, pectoral, dorsal, and caudal fins become reddish. Caudal fin has numerous black bars. Adults reach up to 60 cm in length and up to 4.3 kg (Rakocy, 2014).

Nutritional Facts

Proximate Composition (% WW)

Moisture	Protein	Lipid	Ash
78.325	14.328	1.298	4.953

Source: Data from Ayeloja et al. (2013).

Amino Acids (g/100 g)

Threonine	3.26
Valine	3.69
Methionine	2.00
Isoleucine	3.69
Leucine	5.90
Phenylalanine	3.22
Histidine	1.87
Lysine	7.17
Arginine	4.29
Total EAAs	35.09
Serine	4.22
Glutamic acid	7.04
Proline	2.42
Glycine	3.34
Alanine	4.37
Aspartic acid	6.66
Tyrosine	1.89
Total NEAAs	29.94

EAA, essential amino acid; NEAA, nonessential amino acids.

Source: Data from Taşbozan et al. (2013b).

Minerals (mg/g)

Fe	Cu	Ni	Na	K	Zn	Pb
0.07	0.04	0.032	3.6	9.2	0.073	0.014

Source: Data from Onyia et al. (2010).

Fatty Acids (mg/100 g)

Total lipids: 16:0	208.5
18:1n-9	344.3
18:2n-6	109.6
Linolenic acid (LNA) (18:3n-3)	29.9
Eicopentaenoic acid (EPA) (20:5n-3)	3.3
Docosahexaenoic acid (DHA) (22:6n-3)	12.9
Saturated (SFA)	296.2
Monounsaturated (MUFA)	415.0
Polyunsaturated (PUFA)	175.6
Ratio PUFA:MUFA:SFA	1:2.4:1.7
Ratio omega-6/omega-3 fatty acids	2.8

Source: Data from Petenuci et al. (2008).

Sarotherodon galilaeus Linnaeus 1758

Order: Perciformes

Family: Cichlidae

Common name: Mango tilapia.

Distribution: Africa and Eurasia (Torres, 1991o).

Habitat: Freshwater and brackish water; lakes and coastal rivers.

Description: This species has a deep body and large dorsal and anal fins.

The high dorsal fin rises well forward (about a fifth of the way along the body). Anal, pectoral, and pelvic fins are relatively large and well forward. Tail fin is flexible and fan shaped. Dorsal spines are stout and excavated on alternate sides. This species reaches a maximum size of 34 cm and 1.6 kg (Torres, 1991o; Johnson, 1974).

Nutritional Facts

Proximate Composition (% WW)

Moisture	Protein	Fiber	CHO^a	Ash	Energy^b
70.36	21.52	0.86	2.47	2.20	0.657

Minerals (%)

P	Ca	K	Mg	Fe	Na	Cu	Zn
0.25	0.24	0.34	0.22	0.10	0.73	0.03	0.35

Source: Data from Fawole et al. (2007).

a Carbohydrate.
b MJ/kg DM.
Source: Data from Anene et al. (2013).

Tilapia guineensis Günther 1862

Order: Perciformes

Family: Cichlidae

Common name: Guinean tilapia.

Distribution: Africa: Senegal to Angola (Reyes, 1993k).

Habitat: Freshwaters, brackish waters, and marine waters; rivers.

Description: Upper profile of head of this species is strongly convex. Mouth is terminal. A total of 29–30 scales is seen on lateral line. Outer row teeth are robust. Dorsal fin has 14–16 spines and anal fin has 3 spines. Anal fin of this species is gray and the ventral fins are gray or black and marked by a white line on the anterior edge. Dorsal fin is gray or transparent. Caudal fin is bluish gray and banded with lighter colored spots and a distinctly shaded upper and lower portion. This species grows to a length of 30 cm (Reyes, 1993k).

Sarotherodon melanotheron Rüppell 1852 (= *Tilapia melanotheron*)

Order: Perciformes

Family: Cichlidae

Common name: Blackchin tilapia.

Distribution: Africa: Lagoons and estuaries from Mauritania to Cameroon; several countries in Asia, the United States, and Europe (Torres, 1991p).

192

Habitat: Freshwater, brackish water, estuaries and lagoons; mangrove areas.

Description: In this species, length of caudal peduncle is 0.6–0.9 times its depth. Melanic areas are usually present on lower parts of head, on cleithrum, and on apices of caudal and soft dorsal fins in adults. Occasional irregular and asymmetrical spots are seen on flanks, probably representing vertical bars. The median spot or transverse bar noticed on the nape is rather constant. This species grows to a maximum size of 28 cm (Torres, 1991p).

Nutritional Facts

Proximate Composition (% WW)

	Protein	Lipid	Ash	Moisture
T. guineensis	18.65	0.55	1.30	79.5
T. melanotheron	18.74	0.70	1.06	79.5

Source: Data from Abimbola et al. (2010).

Tilapia rendalli Boulenger 1897

Order: Perciformes

Family: Cichlidae

Common name: Redbreast tilapia.

Distribution: Africa: Kasai drainage (middle Congo River Basin), throughout upper Congo River drainage, Lake Tanganyika, Lake Malawi, Zambesi, coastal areas from Zambesi delta to Natal, Okavango, and Cunene (Bailly, 1997o).

Habitat: Freshwater and brackish water; well-vegetated water along river littorals or backwaters, floodplains, and swamps.

Description: Head and body of this species are mid to dark olive-green dorsally, paling over the flanks. Body has usually vertical bars only, and scales have a dark basal crescent. Dorsal fin is olive-green with a thin red margin, and white to gray dark oblique spots are seen on the soft rays. Caudal fin is spotted on dorsal half and red or yellow on ventral half. This species grows to a maximum size of 45 cm and 25 kg (Bailly, 1997o).

Nutritional Facts

Proximate Composition (%)

Dry Matter	Protein[a]	Lipid[a]	Ash[a]
26.06	20.52	3.52	1.24

[a] Wet weight values.

Amino Acids (g/100 g)

Threonine	3.49
Valine	3.69
Methionine	1.64
Isoleucine	4.08
Leucine	6.10
Phenylalanine	3.24
Histidine	1.86
Lysine	7.10
Arginine	4.91
Total EAAs	36.11

Continued

193

Serine	4.34
Glutamic acid	6.88
Proline	2.79
Glycine	3.76
Alanine	4.57
Aspartic acid	8.27

Tyrosine	1.73
Total NEAAs	32.34

EAA, essential amino acid; NEAA, nonessential amino acid.

Source: Data from Taşbozan et al. (2013b).

Tilapia zillii Gervais 1848

Order: Perciformes

Family: Cichlidae

Common name: Redbelly tilapia.

Distribution: Africa and Eurasia: South Morocco, Sahara, Niger-Benue system, Rivers Senegal, Sassandra, Bandama, Boubo, Mé, Comoé, Bia, Ogun, and Oshun, Volta system, Chad-Shari system (Ortañez, 2005c).

Habitat: Freshwater and brackish water; shallow, vegetated areas.

Description: Upper profile of head of this species is not convex. Dorsal fin has 14–16 spines and 10–14 soft rays. Dark longitudinal band appears on flanks when agitated. No bifurcated dark vertical bars are seen on flanks. Body is brownish olivaceous with an iridescent blue sheen. Lips are bright green and chest is pinkish. Dorsal, caudal, and anal fins are brownish olivaceous with yellow spots. Dorsal and anal fins are outlined by a narrow orange band. This species grows to a maximum size of 40 cm and 300 g (Ortañez, 2005c).

Nutritional Facts

Proximate Composition (% WW)

Protein	Lipid	Moisture	Ash
19.56	0.96	76.75	1.11

Major Amino Acids (mg/g)

Glutamic acid	29.03
Aspartic acid	17.85
Lysine	16.57
Leucine	15.17
Arginine	10.94
Alanine	10.82

Medium Amino Acids (mg/g)

Valine	8.28
Isoleucine	8.05
Glycine	8.31
Threonine	7.67
Serine	7.15
Phenylalanine	6.49
Proline	6.32
Methionine sulfone	5.07

Minor Amino Acids (mg/g)

Histidine	3.97
Cystine	1.84
Tyrosine	2.35
Ornithine	0.43
Taurine	2.41
g-Aminobutyric acid	1.05
Hydroxyproline	0.48

Source: Data from Osibona (2011).

Etroplus suratensis **Bloch 1790**

Order: Perciformes

Family: Cichlidae

Common name: Pearlspot, green chromide.

Distribution: Western Indian Ocean: India and Sri Lanka (Bindu, 2006).

Habitat: Brackish water but known to tolerate freshwater or marine water for short periods: large rivers, reservoirs, lagoons, and estuaries.

Description: Body of this species is deep and laterally compressed. Mouth is small and terminal with a small cleft. Snout is spout-like. Eyes are large and lateral. Lips are thin and jaws are equal. Dorsal fin is inserted above the pectoral fin base. Caudal fin is emarginated, while pelvic fin is characterized with one spine. As in all cichlids, the lateral line is interrupted in this species, too. The lateral line has 6–18 scales after which it continues as a small round hole in each scale. Body color is light greenish with eight yellowish oblique bands. Scales above the lateral line have a central pearly spot and possess some triangular black spots on the abdomen. Fins, except the pectoral, are of a dark leaden color, while the pectoral is yellowish with a jet black base. It grows to a maximum size of 40 cm (Bindu, 2006).

Nutritional Facts

Proximate Composition (% DM)

Moisture	CHO[a]	Protein	Fat	Ash
1.18	20.23	16.55	19.76	1.30

[a] Carbohydrate.

Amino Acids (% DM)

Methionine	Tryptophan
1.55	1.11

Source: Data from Ali et al. (2013).

FAMILY: HELOSTOMATIDAE (KISSING GOURAMI)

Helostoma temminckii **G. Cuvier 1829**

Order: Perciformes

Family: Helostomatidae

Common name: Green kissing fish, pink kissing fish, marbled kissing fish, kissing gourami.

Distribution: Thailand to Indonesia; Southeast Asia (Froese and Pauly, 2007b).

Habitat: Lakes and rivers; shallow, slow-moving, and thickly vegetated backwaters.

195

Description: Body of this species is deep and strongly compressed laterally. Long-based dorsal and anal fins mirror each other in length and frame the body. The posterior-most soft rays of each of these fins are slightly elongated to create a trailing margin. Foremost rays of the jugular pelvic fins are also slightly elongated. Pectoral fins are large, rounded, and low-slung. Caudal fin is rounded to concave. Lateral line is divided in two, with the posterior portion starting below the end of the other. There are three color variations of this fish, viz., a pink or flesh-colored form, a silver-green, and

a mottled variey. It grows to a maximum size of 30 cm (Froese and Pauly, 2007b).

Nutritional Facts

Proximate Composition (g/100 g edible portion, WW)

Energy[a]	Water	Protein	Fat	CHO[b]	Ash
85	78.3	19.8	0.6	0	1.3

[a] kcal.
[b] Carbohydrate.

Minerals (mg/100 g edible portion)

Ca	P	Fe	Na	K
77	241	0.8	18	303

Vitamins (per 100 g edible portion)

Retinol	Carotene	Thiamin	Riboflavin	Niacin	Vitamin C
38 µg	0 µg	0.03 mg	0.46 mg	3.0 mg	3.1 mg

Source: Data from Tee et al. (1989).

FAMILY: OSPHRONEMIDAE (GOURAMIES)

Colisa fasciatus (Bloch and Schneider) (= *Trichogaster fasciata*)

Order: Perciformes

Family: Osphronemidae

Common name: Giant gourami, striped gourami.

Distribution: Asia: Pakistan, India, Nepal, Bangladesh, and upper Myanmar (Torres, 1991d).

Habitat: Streams, pools, beels, ponds.

Description: Body of this species is egg shaped and strongly compressed. Dorsal and abdominal profiles are almost equally convex. Eyes are large. Lips are very thick. Lateral line is interrupted. Scales are present on head,

body, and base of soft dorsal and anal fins. Pectoral fins are thread-like and reach rarely up to the caudal fin. Body color is greenish or black above. Pelvics have yellow-white bases and brilliant red tips. Dorsal and caual fins are spotted with orange. Caudal fin has a cut square appearance (Sultana, 2010b).

Nutritional Facts

Proximate Composition (% WW)

Moisture	Ash	Protein	Lipid
80.75	0.85	15.82	2.58

Source: Data from Ahmed et al. (2012).

Osphronemus goramy Lacepède 1801

Order: Perciformes
Family: Osphronemidae
Common name: Giant gourami, common gourami, true gourami.
Distribution: Native to Southeast Asia; Asia: probably limited to Sumatra, Borneo, Java, the Malay Peninsula, Thailand, and Indochina (Mekong Basin) (Torres, 1991k).
Habitat: Freshwater or brackish water; slow-moving areas such as swamps, lakes, and large rivers.
Description: Juveniles of this species have a pointed snout, a flat head, and an attractive banded coloration. They also have stripes of a silvery blue-gray to black on a cream to golden-yellow background. When they get older, they lose the stripes and turn completely black. Adults develop a swollen forehead (especially the males), along with thick lips (more pronounced on females), and a thick chin. Caudal fin is rounded or obtusely rounded, not truncate or emarginate. Pelvic fins have the first soft ray prolonged into a thread-like tentacle reaching posteriorly to or beyond the hind margin of the caudal fin. This species grows to a maximum size of 70 cm (Torres, 1991k).

Nutritional Facts

Proximate Composition (g/100 g edible portion, WW)

Energy[a]	Water	Protein	Fat	CHO[b]	Ash
110	75.9	19.0	3.8	0.2	1.1

[a] kcal.
[b] Carbohydrate.

Minerals (mg/100 g edible portion)

Ca	P	Fe	Na	K
19	187	0.3	21	328

Vitamins (per 100 g edible portion)

Retinol	Carotene	Thiamin	Riboflavin	Niacin	Vitamin C
47 µg	0 µg	0.05 mg	0.06 mg	2.5 mg	0.7 mg

Source: Data from Tee et al. (1989).

Amino Acids (g/16 g N)

Lysine	8.4
Histidine	2.9
Arginine	7.0
Aspartic acid	7.6
Threonine	3.4
Serine	3.1
Glutamic acid	14.7
Proline	4.7
Glycine	3.7
Alanine	5.0
Cystine	0.8
Valine	4.0
Methionine	2.7
Leucine	7.3
Isoleucine	4.2
Tyrosine	3.5
Phenylalanine	4.0

Source: Data from Zanariah and Rehan (1988).

197

Trichogaster pectoralis Regan 1910

Order: Perciformes

Family: Osphronemidae

Common name: Snakeskin gourami.

Distribution: Cambodia, Thailand, southern Vietnam, and Laos; Philippines, Malaysia, Indonesia, Singapore, Papua New Guinea, Sri Lanka, and New Caledonia.

Habitat: Rice paddies, shallow ponds, and swamps.

Description: It is an elongated, moderately compressed fish with a small dorsal fin. Its anal fin is nearly the length of the body, and the pelvic fins are long and thread-like. Back is olive in color and the flanks are greenish gray with a silver iridescence. A clear, irregular black band extends from the snout, through the eye, and to the caudal peduncle. Underparts are white. Rear part of the body may be marked with faint transverse stripes. Fins are gray-green. Dorsal fins of male fish are pointed, and the pelvic fins are orange to red. The males are slimmer than the less colorful females. Juveniles have strikingly strong zigzag lines from the eye to the base of the tail. This species can grow up to 25 cm.

Trichogaster trichopterus Pallas 1770

Order: Perciformes

Family: Osphronemidae

Common name: Three-spot gourami, blue gourami.

Distribution: Native to southeastern Asia; endemic to the Mekong Basin in Cambodia, Laos, Thailand, and Vietnam, and Yunnan in Southeast Asia.

Habitat: Marshes, swamps, canals, and lowland wetlands.

Description: This species gets its name from the two spots along each side of its body in line with the eye, which is considered the third spot. Median fins and pectorals are brown, and ventrals are yellowish. Mouth is very small and very oblique. Upper jaw is vertical and somewhat protractile and lower jaw is prominent. Lateral line is curved and irregular. Caudal fin is slightly emarginate or truncate. Color when alive is brown. Shoulders possess irregular dark marks. Opercles and thorax are yellowish. A black spot is seen in the middle of the side and at the caudal fin base. Body has numerous narrow irregular oblique bars. Blue gouramis are among those fish that possess a labyrinth organ, which allows them to breath air directly. It reaches a maximum length of 15 cm (Capuli, 1991).

Nutritional Facts

Proximate Composition (% WW)

	Moisture	Protein	Fat	Ash	CHO[a]
T. pectoralis	80.0	16.0	1.5	1.0	1.5
T. trichopterus	78.0	17.5	2.2	1.0	1.3

[a] Carbohydrate.

Amino Acids (g/16 g N)

	T. pectoralis	*T. trichopterus*
Lysine	6.9	10.6
Histidine	0.9	3.5
Arginine	6.1	8.5
Aspartic acid	4.6	9.0
Threonine	4.6	4.6
Serine	4.0	4.0
Glutamic acid	18.8	14.6
Proline	2.8	4.6
Glycine	3.5	5.4
Alanine	6.3	5.9

Cystine	0.8	0.9
Valine	4.8	5.7
Methionine	3.1	2.9
Leucine	9.1	8.2
Isoleucine	5.7	3.7
Tyrosine	4.0	3.3
Phenylalanine	4.3	3.8

Source: Data from Zanariah and Rehan (1988).

Minerals and Vitamins of *T. pectoralis*
Minerals (mg/100 g Edible Portion)

Ca	P	Fe	Na	K
62	210	0.7	10	282

Vitamins (per 100 g Edible Portion)

Retinol (µg)	Carotene (µg)	Thiamin (mg)	Riboflavin (mg)	Niacin (mg)	Vitamin C (mg)
25	0	0.12	0.36	1.8	2.6

Source: Data from Tee et al. (1989).

Trichogaster lalius F. Hamilton 1822 (= *Colisa lalia*)

Order: Perciformes
Family: Osphronemidae

Common name: Dwarf gourami.

Distribution: Originating from India, West Bengal, Assam, and Bangladesh (About home).

Habitat: River plains; slow-moving waters in rivulets, streams, and lakes with plentiful vegetation.

Description: Males of this species are slightly larger than the females and have a bright orange-red body with turquoise blue vertical stripes that extend into the fins. Dorsal fin of the male is pointed in contrast to the rounded dorsal of the female. Females remain a duller silvery blue-gray color, never achieving the brilliant colors of the male. Many color hybrids exist, including blue–powder blue), neon, rainbow, and red-blushing. This species reaches a maximum length of only 5 cm (About home).

Nutritional Facts

Proximate Composition (% WW)

Moisture	Ash	Protein	Lipid
77.52	2.21	16.13	4.15

Source: Data from Ahmed et al. (2012).

FAMILY: PERCIDAE (PERCHES)

Perca flavescens Mitchill 1814

Order: Perciformes

Family: Percidae

Common name: Yellow perch.

Distribution: Only found in North America.

Habitat: Large and small lakes; slow-moving rivers, streams, brackish waters, and ponds.

Description: Body of this species is laterally compressed. Anterior portion of the body is deep, gradually tapering into a slender caudal peduncle. Opercle is partially scaled, and a single spine is present on the posterior margin. Caudal fin is forked. This species has a yellow to brass-colored body and distinct pattern, consisting of five to nine olive-green vertical bars, triangular in shape, on each side. Its fins are lighter in coloration, with an orange hue on the margins. This species reaches a common length of 35 cm and 1 kg.

Nutritional Facts

Proximate Composition (% DM)

Moisture[a]	Lipid[b]	Protein[b]
81.3	1.39	94.3

[a] Wet weight.
[b] Freeze-dried.

Amino Acids (g/100 g of freeze-dried DM)

Alanine	4.19
Arginine	5.46
Aspartic acid	7.48
Cysteine	0.26

Continued

200

Glutamic acid	11.60
Glycine	3.00
Histidine	1.77
Isoleucine	4.28
Leucine	6.59
Lysine	8.12
Methionine	2.58
Phenylalanine	3.57
Proline	2.41

Serine	1.92
Threonine	2.68
Tyrosine	2.72
Valine	4.10

Minerals (mg/100 g DM)

Fe	Cu	Mn	Zn	Na	Mg	P	K	Ca
4.65	0.38	0.16	7.24	0.50	0.22	1.64	2.63	0.16

Source: Data from González et al. (2006).

Perca fluviatilis Linnaeus 1758

Order: Perciformes

Family: Percidae

Common name: European perch.

Distribution: Eurasia (Torres, 1991l).

Habitat: Slow-flowing rivers, deep lakes, and ponds.

Description: Body of this species is moderately deep and covered with rough-edged scales. Head is short and snout rounded and blunt. There are two dorsal fins, clearly separated from each other. First dorsal fin is markedly higher than the second. Caudal fin is is emarginated. A total of 56–77 scales are seen along a lateral line. Body is greenish yellow with five to nine transverse black bands on the sides. First dorsal fin is gray with a black spot at the tip. Second dorsal is greenish yellow and pectorals are yellow. Other fins are red. It attains a maximum length of 51 cm and weight of 4.75 kg (Torres, 1991l).

Nutritional Facts

Proximate Composition (% WW)

Moisture	Protein	Fat	Ash
80.9	17.6	0.3	1.2

Source: Data from Jankowska et al., 2007.

Lipid Content of the Meat

% of Wet Weight	% of Dry Matter
0.8	4

201

Major Elements in the Muscle (mg/100 g)

Na	K	Mg	Ca	P
47	330	20	20	198

Trace Elements in the Muscle (µg/100 g)

Fe	Mn	Se
1000	0	24

Vitamins in the Muscle (µg/100 g)

A	B1	B2	B6
7	175	120	0

Source: Data from Steffens (2006).

Sander lucioperca Linnaeus 1758

Order: Perciformes

Family: Percidae

Common name: Pike-perch.

Distribution: Europe and Asia: Caspian, Baltic, Black, and Aral Sea Basins; Elbe (North Sea Basin) and Maritza (Aegean Basin) (Luna, 1988r).

Habitat: Freshwater and brackish water; large, turbid rivers and eutrophic lakes, brackish coastal lakes and estuaries.

Description: Body of this species is elongated. Snout is pointed and head length is greater than depth of body or equal to it. Upper jaw extends past eye level. There are two dorsal fins; the first one is spiny and separated by a narrow interspace from the second. Anal fin has two to three spines. Pelvic fins are widely spaced; the distance between them is almost as great as the base of one fin. Lateral line is with 84–95 scales. Color of body is greenish gray or brown on the back, becoming lighter on the lower sides and white on the belly. This species grows to a maximum size of 1 m and 20 kg (FAO, 2014g; Luna, 1988r).

Nutritional Facts

Proximate Composition (g/100 g DM)

Protein	Fat	Fiber	Ash	Energy[a]
51.25	17.44	1.90	9.31	22.52

[a] MJ/kg.

Amino Acids (g/100 g protein)

Arginine	6.31
Histidine	4.11
Lysine	8.64
Tryptophan	0.88
Phenylalanine	5.54
Tyrosine	5.04
Methionine	3.41
Cysteine	3.27
Threonine	5.35
Leucine	9.47
Isoleucine	6.50
Valine	6.69
Alanine	6.53
Glycine	6.42
Glutamic acid	18.45
Asparagine	10.65
Proline	5.75
Serine	5.63

Source: Data from Jarmolowicz and Zakês (2014).

Protein Content of the Muscle

% of Wet Weight	% of Dry Matter
19.2	89

Lipid Content of the Muscle

% of Wet Weight	% of Dry Matter
0.7	3

Major Elements in the Muscle (mg/100 g)

Na	K	Mg	Ca	P
24	377	50	49	191

Trace Elements in the Muscle (µg/100 g)

Fe	Mn	Se
754	90	26

Vitamins in the Muscle (µg/100 g)

A	B1	B2	B6
0	160	250	0

Source: Data from Steffens (2006).

FAMILY: POLYNEMIDAE (THREADFINS)

Pentanemus quinquarius Linnaeus 1758

Order: Perciformes

Family: Polynemidae

Common name: Royal threadfin.

Distribution: Eastern Atlantic: Senegal to Angola; Cuba (Garilao, 1995a).

Habitat: Marine water and brackish water; sandy and muddy bottoms in shallow waters, frequently in brackish water habitats.

Description: Body of this species is elongated and fusiform, with spinous and soft dorsal fins widely separated. Tail fin is large and deeply forked. Mouth is large and inferior. A blunt snout projects far ahead. Its most distinguishing feature is its pectoral fins, which are composed of two distinct sections, the lower of which consists of three to seven long,

203

thread-like independent rays that may extend well past the tail fin. This species grows to a maximum size of 35 cm.

Nutritional Facts

Proximate Composition (% WW)

Protein	Lipid	Moisture	Ash
19.80	0.40	72.63	1.13

Major Amino Acids (mg/g)

Glutamic acid	30.77
Aspartic acid	19.21
Lysine	17.60
Leucine	16.05
Arginine	11.36
Alanine	10.95

Medium Amino Acids (mg/g)

Valine	8.50
Isoleucine	8.48
Glycine	7.20
Threonine	7.99
Serine	7.64
Phenylalanine	7.42
Proline	5.69
Methionine sulfone	5.60

Minor Amino Acids (mg/g)

Histidine	3.71
Cystine	1.99
Tyrosine	1.25
Ornithine	0.34
Taurine	3.20
g-Aminobutyric acid	1.25
Hydroxyproline	ND

ND, not detected.

Source: Data from Osibona (2011).

FAMILY: POMATONIDAE (BLUE FISHES)

Pomatomas saltatrix Linnaeus 1766

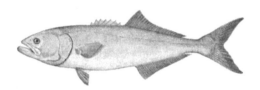

Order: Perciformes

Family: Pomatomidae

Common name: Bluefish.

Distribution: Subcosmopolitan in tropical and subtropical seas. Western and eastern Atlantic Ocean. Mediterranean and Black Seas. Western and eastern Indian Ocean. Western central Pacific (FAO, 2014e).

Habitat: Estuaries and seas.

Description: Body of this species is elongated and compressed. Head is large and keeled above. Scales are small, covering head and body and bases of fins. Lateral line is almost straight. Mouth is terminal and lower jaw is sometimes slightly prominent. There are two dorsal fins: the first is short and low, with seven or eight feeble spines connected by a membrane, and the second is long. Pectoral fins are short, not reaching to origin of soft dorsal fin. Anal fin is a little shorter than the soft dorsal fin with two spines. Caudal fin is forked, but not deeply so. Color of back is greenish blue and sides and belly are silvery. Dorsal and anal fins are pale green tinged with yellow.

Pectoral fins are bluish at base. Caudal fin is dull greenish tinged with yellow. Maximum values of length and weight observed are 120 cm and 14 kg, respectively (FAO, 2014e).

Nutritional Facts

Proximate Composition (% WW)

Moisture	Ash	Protein	Fat
70.16	1.12	19.56	2

Source: Data from Sarma et al. (2011).

Nutrition Facts (per 3 oz WW)

	Amount per Serving
Calories	105.4 kcal
Total fat	3.6 g
Saturated fat	0.8 g
Polyunsaturated fat	0.9 g
Monounsaturated fat	1.5 g
Cholesterol	50.2 mg
Sodium	51 mg
Potassium	316.2 mg

Carbohydrate	0 g
Dietary fiber	0 g
Protein	17 g
Vitamin A	338.3 IU
Vitamin C	0 mg
Thiamin	0 mg
Riboflavin	0.1 mg
Niacin	5.1 mg
Vitamin B6	0.3 mg
Folate	1.7 µg
Folic acid	0 µg
Vitamin B12	4.6 µg
Pantothenic acid	0.7 mg
Calcium	6 mg
Iron	0.4 mg
Magnesium	28.1 mg
Zinc	0.7 mg
Copper	0 mg
Phosphorus	193 mg
Selenium	31 µg

Source: Data from U.S. Department of Agriculture, Standard Release 18.

FAMILY: NANDIDAE (ASIAN LEAF FISHES)

Nandus nandus Hamilton 1822

Order: Perciformes

Family: Nandidae

Common name: Gangetic leaf fish.

Distribution: Asia: Pakistan to Thailand (Bailly, 1997h).

Habitat: Freshwater and brackish water; ditches and inundated fields; dried-up beds of tanks, beels, bheries; standing or sluggish waters of lakes, reservoirs, or canals.

Description: It is an oval-shaped fish with an arched back and lateral compression. First 12 rays of the long dorsal fin are spiny, while the rest are not. Caudal fin is fan shaped and the

205

mouth is deeply cleft. Body is generally gray with irregular brown markings. Eye has two brown stripes passing through it: one running from the mouth to the origin of the dorsal fin, and the other running from the throat to the eye. Fins are grayish with brown markings. It grows to a maximum size of 20 cm (Mongabay.com).

Nutritional Facts

Proximate Composition (% WW)

Moisture	Ash	Protein	Lipid
78.61	2.83	15.80	2.75

Source: Data from Ahmed et al. (2012).

FAMILY: SCOMBRIDAE (MACKERELS)

Scoberomorus tritor Cuvier 1832

Order: Perciformes

Family: Scombridae

Common name: West African Spanish mackerel.

Distribution: Eastern Atlantic: Canary Islands and Senegal to the Gulf of Guinea and Baía dos Tigres, Angola (Kesner-Reyes, 2002c).

Habitat: Marine and brackish water; estuaries.

Description: Interpelvic process of this species is small and bifid. Body is covered with small scales. Lateral line is gradually curving down toward caudal peduncle. Swim bladder is absent. Large individuals may have thin vertical bars. Coloration of body is bluish green on the back, fading to silvery on the sides marked with about three rows of vertically elongate orange spots. Anterior half of first dorsal fin and margin of posterior half of first fin are black. This species grows to a maximum size of 100 cm and 6 kg (Kesner-Reyes, 2002c).

Nutritional Facts

Proximate Composition (% WW)

Moisture	Protein	Fiber	CHO[a]	Ash	Energy[b]
69.52	21.52	0.76	3.51	2.50	0.669

[a] Carbohydrate.
[b] MJ/kg DM.

Source: Data from Anene et al. (2013).

ORDER: SILURIFORMES (CATFISHES)

FAMILY: SILURIDAE (SHEATFISHES)

Ompok bimaculatus Bloch 1794

Order: Siluriformes

Family: Siluridae

Common name: Butter catfish.

Distribution: Asia: Indian subcontinent and Myanmar (Reyes, 1993i).

Habitat: Freshwater, brackish water; quiet, shallow (0.5–1.5 m), often muddy water, in sandy streams, rivers, and tanks; canals, beels, and inundated fields.

Description: Body of this species is elongated and compressed. Head is small and broad. Snout is bluntly rounded. Mouth is superior. Lips are thin. Jaws are subequal and lower jaw is prominent. Barbels are present. Dorsal fin is inserted above last half of pectoral fin. Pelvic fins are short. Anal fin is very long and close to caudal fin, which is forked. Lateral line is complete. This species grows to a maximum size of 45 cm (Shukla, 2005).

Nutritional Facts

Proximate Composition (% DM)

Moisture	Protein	Lipid	Ash
15.54	30.02	19.63	5.33

Macroelements (mg/100 g)

Ca	Mg	K	Na	P
24.25	106.75	121.05	103.12	483.33

Microelements (mg/100 g)

Cu	Co	Mn	Zn	Fe	Ni	Cr
ND	4.868	0.1315	0.921	1.7105	2.500	1.115

Source: Data from Hei and Sarojnalini (2012).

Kryptopterus micronema Bleeker 1846 (= *Cryptopterus micronema*)

Order: Siluriformes

Family: Siluridae

Common name: Sheatfish.

Distribution: Thailand to Indonesia (Jenkins et al., 2009).

Habitat: Rivers, streams, and lakes as well as impoundments.

Description: Dorsal profile is convex. Snout has a pronounced concavity at the nape. Mouth is large. Maxillary barbels reach the eyes and mandibular barbels are very delicate. Anal fin is free from the caudal fin. Teeth in jaws are rather sharp and curved. Color of the fish is brownish or yellowish and darker on the back. Fins may be powdered with black. Maximum length of the fish is 77 cm (Rainboth, 1996).

Nutritional Facts

Proximate Composition (% DM)

Moisture	Protein	Fat	Ash	CHO[a]
19.91	38.81	8.02	0.73	26.53

[a] Carbohydrate.

Amino Acids (mg/100 g)	
Aspartic acid	5.86
Glutamic acid	9.18
Serine	4.21
Glycine	4.55
Histidine	3.03
Arginine	8.01
Threonine	5.81
Alanine	4.10
Proline	2.81
Tyrosine	2.44
Valine	6.34
Methionine	2.26
Cystine	1.84
Isoleucine	4.63
Leucine	6.92
Phenylalanine	5.19
Lysine	5.80

Source: Data from Hudu et al. (2010).

Silurus triostegus Heckel 1843

Order: Siluriformes

Family: Siluridae

Common name: Planet catfish.

Distribution: Asia: Iraq (Binohlan, 1990c).

Habitat: Freshwater, brackish water; lakes, rivers.

Description: Mouth of this species is large. Lower jaw is longer than upper jaw. Maxillary barbels, which are longer than the head, reach the end of the head. Mandibular barbels, which are shorter, reach only the middle. Anal fin is not fused to the caudal fin, which is rounded. The upper part of the body is mottled pale yellow-brown and black. Belly has black spots. This species reaches a maximum size of 99 cm and 8.5 kg (Binohlan, 1990c; Unlu et al., 2012).

208

Nutritional Facts

Proximate Composition (g/100 g WW)

Protein	Lipid	Moisture	Ash	Energy[a]
17.68	4.87	76.56	0.84	114.75

[a] kcal/100 g.

Minerals (µg/g WW)

Ca	P	K	Mg
92.59	1447.56	2762.50	227.26

Source: Data from Olgunoglu et al. (2014).

Silurus glanis Linnaeus 1758

Order: Siluriformes

Family: Siluridae

Common name: Wels catfish.

Distribution: Europe and Asia; North, Baltic, Black, Caspian, and Aral Sea Basins; as far north as southern Sweden and Finland (Torres, 1991s).

Habitat: Freshwater and brackish water; large and medium-sized lowland rivers, backwaters, and well-vegetated lakes.

Description: This species has two pairs of mental barbels and its anal fin has 83–95 rays. Body is naked. Head is large and depressed. Dorsal fin has 2–5 rays and caudal fin is rounded or truncate. There is no adipose fin. Anal rays are almost touching the caudal fin. Caudal fin has 17 rays. This species grows to a maximum length of 500 cm and weight of 306 kg (Torres, 1991s).

Nutritional Facts

Protein Content of the Meat

% of Wet Weight	% of Dry Matter
15.3	55

Lipid Content of the Meat

% of Wet Weight	% of Dry Matter
11.3	41

n-3 and n-6 Fatty Acids (%) in the Lipids

Σn-3	Σn-6	n-3/n-6
20.1	11.1	1.8

Major Elements in the Muscle (mg/100 g)

Na	K	Mg	Ca	P
20	421	55	27	151

Trace Elements in the Muscle (µg/100 g)

Fe	Mn	Se
600	79	0

Source: Data from Steffens et al. (2006).

Wallago attu Bloch & Schneider 1801

Order: Siluriformes

Family: Siluridae

Common name: Wallago.

Distribution: Asia: Pakistan to Vietnam and Indonesia; Afghanistan (Torres, 1991v).

Habitat: Freshwater, brackish water; large rivers, lakes, tanks.

Description: Body of this species is elongated and strongly compressed. Head is broad and snout is depressed. Mouth is very deeply cleft and its corner reaches far behind the eyes. Teeth in jaws are set in wide bands. Barbels are in two pairs. Maxillary barbels extend to anterior margin of the anal fin. Mandibular barbels, which are longer than the pelvic fin, extend to the angle of the mouth. Eyes are small with a free orbital margin. Dorsal fin is small and anal fin is very long. Eyes are located in front of the vertical through the corner of the mouth. This species reaches a maximum size of 240 cm (Torres, 1991v).

Nutritional Facts

Proximate Composition (% WW)

	Moisture	Ash	Fat	Protein
Male	76.19	3.36	4.03	15.69
Female	76.0	3.31	4.27	15.69

Source: Yousaf et al. (2011)

FAMILY: SCHILBEIDAE (SCHILBID CATFISHES)

Eutropiichthys vacha Hamilton 1822

Order: Siluriformes

Family: Schilbeidae

Common name: Batchwa bacha.

Distribution: Asia: Pakistan, India, Bangladesh, Nepal, Myanmar, and Thailand (Luna, 1988i).

Habitat: Freshwater, brackish water; rivers, canals, and tidal waters.

Description: Body of this species is elongated and laterally compressed. Dorsal and ventral profiles are almost convex. Upper jaw is longer than lower jaw. Four pairs of barbels are present. Dorsal spine is serrated posteriorly, and pectoral is serrated internally. Adipose fin is present. Caudal fin is deeply forked. Body is silver colored with grayish back. Pectoral and anal fins have reddish margin. Lateral line is complete. It grows to a maximum size of 34 cm and 1.4 kg (Nayan, 2011d).

Proximate Composition (% DM)

Moisture	CHO[a]	Protein	Fat	Ash
2.20	19.50	17.34	35.00	1.10

[a] Carbohydrate.

Amino Acids (% DM)

Methionine	Tryptophan
1.20	1.30

Source: Data from Ali et al. (2013).

Ailia coila **Hamilton 1822**

Order: Siluriformes

Family: Schilbeidae

Common name: Gangetic ailia.

Distribution: Asia: Pakistan, India, Bangladesh, and Nepal (Luna, 1988b).

Habitat: Freshwater, brackish water; large rivers and connected waters.

Description: Body of this species is elongated and deeply compressed. Mouth is subinferior. Upper jaw is longer than lower jaw. Barbels are well developed and are in four pairs. Dorsal fin is absent. A small adipose fin is present. Pectorals are well developed and pelvic fin is small. Pectoral spine is slender and serrated. Anal fin is large. Caudal fin is forked, and its lower lobe is longer than the upper lobe. Body is silvery or whitish in color. This species grows to a maximum size of 30 cm (Nayan, 2011a).

Nutritional Facts

Proximate Composition (% WW)

Protein	Fat	Moisture	Ash
16.99	3.53	78.62	1.98

Source: Data from Mazumder et al. (2008).

Clupisoma atherinoides **Bloch 1794 (=** *Pseudeutropius atherinoides, Neotropius atherinoides*)

Order: Siluriformes

Family: Schilbeidae

Common name: Indian potasi.

211

Distribution: Asia: Pakistan, India, Bangladesh, Nepal, and Myanmar (Talwar and Jhingran, 1991).

Habitat: Freshwater, brackish water, tidal rivers.

Description: Body of this species is elongated and compressed. Upper jaw is slightly longer than the lower jaw.

Eyes are large. Body is whitish transparent with a light greenish back. A total of three or four longitudinal bands are present on the flank. Caudal fin is forked and caudal base has a black spot. Barbels are in four pairs. Dorsal spine is long and serrate. This species grows to a maximum size of only 15 cm (Galib, 2008).

Ailiichthys punctata Day 1872 (= *Ailia punctatus*)

Order: Siluriformes

Family: Schilbeidae

Common name: Jamuna ailia.

Distribution: Asia: Pakistan, India, Bangladesh (Rahman, 1989).

Habitat: Large rivers with turbid water and a substrate of sand or mud.

Description: Body of this species is elongated and deeply compressed. Mouth is subinferior and upper jaw is slightly longer than lower jaw. Upper surface of head is covered with thin layer of skin and upper region is slightly concave. A narrow adipose lid is present in each eye. Barbels are in pairs. Dorsal and pelvic fins are absent. Caudal fin is forked, and lower lobe is slightly longer than upper lobe. Body is silvery. Upper surface of head is faint to black. A large black spot is present in the upper edge of the dorsal fin, and the anal fin base is yellowish. This species grows to a maximum size of only 10 cm (Nayan, 2011b).

Nutritional Facts

Proximate Composition (% WW)

	Moisture	Protein	Fat	Ash
A. punctata	75.46	18.43	4.50	1.54
C. atherinoides	76.60	16.69	1.87	2.87

Source: Data from Begum and Minar (2012).

Clupisoma garua Hamilton 1822

Order: Siluriformes

Family: Schilbeidae

Common name: Garua bachcha.

Distribution: Asia: Pakistan, India, Bangladesh, and Nepal (Luna, 1988g).

Habitat: Tidal rivers.

Description: Body of this species is elongated and laterally compressed. Eyes have eyelids. Mouth is moderate and subterminal. Upper jaw is longer than lower jaw. Barbels are in four pairs. Gill opening is wide. Dorsal spine is slender and serrated posteriorly. Body is dark on back and whitish or silvery at sides and abdomen. This species has been reported to reach a maximum size of 60.9 cm (Nayan, 2011c).

Nutritional Facts

Proximate Composition (per 100 g DM)

Protein (g)	Lipid (mg)	Moisture[a] (%)
18.50	1.60	80.00

[a] Wet weight.

Source: Data from Patole (2012).

Sperata aor Hamilton 1822 (= *Macrones aor*)

Order: Siluriformes

Family: Schilbeidae

Common name: Long-whiskered catfish, Indian shovelnose catfish.

Distribution: Asia: Pakistan, India, Nepal, Bangladesh, and upper Myanmar (Luna, 1988t).

Habitat: Rivers, ponds, lakes, channels, and reservoirs.

Description: Body of this species is elongated; head is compressed and mouth is subterminal. Eyes are transversely oval. There are four pairs of barbels, and the maxillary pair reaches to the base of the caudal. Dorsal and pectoral fins have a strong spine that is finely serrated on its posterior edge. The adipose fin, which is well developed, has a dark spot at its posterior base. Caudal fin is forked and its upper lobe is slightly longer than its lower lobe. Lateral line is complete. Males are more slender and have an elongate genital papilla in front of the anal fin. It reaches a maximum length of 180 cm (Nayan, 2011e; PlanetCatfish.com).

Nutritional Facts

Proximate Composition (% WW)

Moisture	Fat	Protein	Ash
76.0–82.5	0.21–4.55	14.9–19.0	0.81–1.45

Source: Data from Natarajan and Srinivasan (1961).

FAMILY: BAGRIDAE (BAGRID CATFISHES)

Sperata seenghala Sykes 1839

Order: Siluriformes

Family: Bagridae

Common name: Giant river catfish.

Distribution: Asia: Afghanistan, Pakistan, India, Nepal and Bangladesh, Thailand, China (Torres, 1991t).

Habitat: Freshwater, brackish water; rivers, canals, beels, ditches, inundated fields, and other freshwater areas.

Description: Body of this species is elongated and compressed. Snout is broad and spatulate. Barbels extend posteriorly to pelvic fins or beyond to anal fin. Dorsal spine is weakly serrated on its posterior edge. Adipose fin base is short, about as long as the rayed dorsal fin base. Color is brownish gray on back and silvery on flanks and belly. A dark well-defined spot is seen on the adipose dorsal fin. This species grows to a maximum size of 150 cm (Torres, 1991t).

Nutritional Facts

Proximate Composition (% WW)

Moisture	Protein	Fat	Ash
79.40	20.06	1.40	0.90

Source: Data from Mohanty et al. (2012).

Bagrus bayad Forsskål 1775

Order: Siluriformes

Family: Bagridae

Common name: Bayad.

Distribution: Africa.

Habitat: Lakes and rivers.

Description: Body of bayad is more or less elongated. Dorsal fin has a smooth spine, and pectoral fins have spines with serrations on the inside. There are four pairs of barbels. Maxillary barbels may reach to the ventral fin or pelvic fins. This fish is yellow greenish or blackish with a white belly. Fins are darker, sometimes reddish purple. Juveniles have little black spots on the sides. This species has a maximum size of 112 cm and weight of 100 kg.

Nutritional Facts

Proximate Composition (g/100 g DM)

Lipid	Protein	Moisture[a]
13.2	77.0	76.0

[a] Wet weight.

Minerals (mg/g DM)

K	P	Ca	Na	Mg	Zn[a]	Fe[a]	Cu[a]
12,100	7730	2920	2360	748	51	17	1.0

[a] µg.

Essential Amino Acids (EAAs) (µg/g DM)

Leucine	422.1
Isoleucine	250.9
Valine	266.8
Threonine	246.6
Lysine	486.4
Histidine	173.6
Methionine	189.9
Phenylalanine	243.1
Total EAAs	2279.4

Source: Data from Mohamed et al. (2010).

Rita rita Hamilton 1822

Order: Siluriformes

Family: Bagridae

Common name: Rita.

Distribution: Asia: Afghanistan, Pakistan, India, Nepal, Bangladesh, and Myanmar.

Habitat: Freshwater, brackish water; rivers and estuaries; muddy to clear water; backwater of quiet eddies.

Description: This species is distinguished by its small eyes and purplish gray color. A single pair of mandibular barbels is present. Top of head is covered by thin skin, and vomerine teeth are seen in two broad ovoid patches. Males have an elongate genital papilla in front of the anal fin. It reaches a maximum size of 150 cm (PlanetCatfish).

Nutritional Facts

Proximate Composition and Mineral (% DM)

Moisture[a]	Fat	Protein	Fiber	Ash	P
68.5	21.3	54.0	7.3	8.9	1.5

[a] Wet weight.
Source: Data from Abbas et al. (2013).

Mystus tengara Hamilton 1822

215

Order: Siluriformes

Family: Bagridae

Common name: Golden catfish, Guinea catfish, Pyjama catfish.

Distribution: Asia: Pakistan, India, Nepal and Bangladesh, Afghanistan (Reyes, 1993h).

Habitat: Flowing and standing waters; rivers and ponds in plains and submontane regions.

Description: Dorsal profile of this species is rising evenly from tip of snout to origin of dorsal fin. Ventral profile is roughly straight up to the end of the anal fin base. Eyes are rounded and located entirely in the dorsal half of the head. Mouth is subterminal. Barbels are in four pairs. Dorsal fin has spinelet, spine, and seven branched rays. The spine is serrated anteriorly near the distal tip. Adipose fin is long. Pectoral fin has a backwardly curved stout spine. Caudal fin is deeply forked. Body has a distinct oval dark brown tympanic spot and four brown stripes (a middorsal and three lateral stripes), and all the stripes are separated by pale longitudinal lines of equal width. This species grows to a maximum size of only 18 cm (Darshan et al., 2013).

Proximate Composition (% WW)

Moisture	Ash	Protein	Lipid
77.17	1.48	17.86	3.48

Source: Data from Ahmed et al. (2012).

Mystus cavasius Hamilton 1822

Order: Siluriformes

Family: Bagridae

Common name: Gangetic mystus.

Distribution: Asia: Indian subcontinent (Pakistan, Nepal, India, Sri Lanka, and Myanmar) (Susan, 1988b).

Habitat: Freshwater, brackish water; tidal rivers and lakes; beels, canals, ditches, ponds, and inundated fields.

Description: Body of this species is elongated and compressed. Head is conical. Maxillary barbels in adults extend posteriorly beyond the caudal fin base. Color is grayish with a more or less well-defined midlateral longitudinal stripe. A dark spot emphasized by a pale area is seen along its ventral margin. Dorsal, adipose, and caudal fins are shaded with melanophores. It grows to a maximum seize of 40 cm and 10 kg (Susan, 1988b).

Proximate Composition (% WW)

Moisture	Ash	Protein	Lipid
78.62	1.27	17.30	2.81

Source: Data from Ahmed et al. (2012).

FAMILY: CLAROTEIDAE (CLAROTEID CATFISHES)

Chrysichthys nigrodigitatus Lacepède 1803

Order: Siluriformes

Family: Claroteidae

Common name: Bagrid catfish, mudfish.

Distribution: Mauritania to Angola (Azeroual et al., 2010)

Habitat: Shallow waters of lakes; rivers, swamps.

Description: Head of this species is oval and eyes are very large and easily visible from above. Dorsal fin is large and its upper edge is round. Caudal fin is deeply forked. Color of body is gray-blue except for the ventral surface, which is white. Fins are grayish pink; adipose fins are black and lips and barbels are pink. This species grows to a maximum size of 65 cm (Reed et al., 1967).

Nutritional Facts

Proximate Composition (% WW)

Moisture	Protein	Fat	CHO[a]	Ash
79.53	21.37	2.13	2.21	3.30

[a] Carbohydrate.

Gross Calorific Values (kcal/100 g)

Protein	Fat	CHO[a]	Total
13.27	27.52	1.35	42.14

[a] Carbohydrate.

Source: Data from Keremah and Amakiri (2013).

Auchenoglanis biscutatus Geoffroy Saint-Hilaire 1809

Order: Siluriformes

Order: Claroteidae

Common name: Black-spotted catfish.

Distribution: Western and northeast Africa (Azeroual et al., 2010)

Habitat: Running water, lakes, and submerged vegetation.

Description: Body and adipose fin in adults of this species are either light brown or gray with small, distinct spots on the rayed fins that do not form bands. Shape of the nuchal plate is also distinctive from all other species in this genus. It feeds on the bottom on mud, debris, and insects, notably chironomid larvae and molluscs. This species reaches a maximum size of 54 cm (Scotcat.com, 2012).

Nutritional Facts

Proximate Composition (% DM)

Moisture	Ash	Protein	Fat	EE	NFE
8.02	23.06	12.93	7.90	36.07	8.96

EE, ether extract; NFE, nitrogen-free extract.
Source: Data from Olele (2012).

FAMILY: CLARIIDAE (AIRBREATHING CATFISHES)

Clarias anguillaris Linnaeus 1758

Order: Siluriformes

Family: Clariidae

Common name: Mudfish.

Distribution: Native of Africa; Togo, Ghana, Gambia, Sénégal, and Algeria (Teugels, 1986).

Habitat: Inundated freshwater areas.

Description: Body of this species is elongated. Head is large, depressed, and bony with small eyes. Gill openings are wide. Air-breathing labyrinthic organ arises from gill arches. Mouth is terminal and large. Four pairs of barbels are present. Dorsal and anal fins are long. Dorsal fin spine and adipose fin are absent. Anterior edge of pectoral spine is serrated. Caudal fin is rounded. Color varies from sandy-yellow through gray to olive with dark greenish brown markings. Belly is white. This species grows to a maximum size of 1 m and 7 kg (Teugels, 1986).

Minerals (mg/g)

Ca	K	Mg	P
2.91	0.78	0.32	0.29

Amino Acids (mg/g)

Alanine	6.05
Arginine	6.74
Aspartic acid	9.89
Glutamic acid	15.68
Glycine	9.98
Histidine	3.49
Isoleucine	3.01
Leucine	7.68
Lysine	6.33
Methionine	2.31
Phenylalanine	4.68
Proline	5.39
Serine	5.18

Source: Data from Effiong and Mohammed (2008).

Nutritional Facts

Proximate Composition (% FW)

Moisture	Ash	Fat	Protein	Fiber
80.85	0.90	6.85	19.22	0.60

Clarias batrachus Linnaeus 1758

Order: Siluriformes

Family: Clariidae

Common name: Philippine catfish.

Distribution: Asia: Java, Indonesia (Luna, 1988f).

Habitat: Freshwater, brackish water; lowland streams, swamps, ponds, ditches, rice paddies, and pools left in low spots.

Description: Body of this species is compressed posteriorly. Upper jaw is a little projecting. Spine of pectoral fins is rough on its outer edge and serrated on its inner edge. Occipital process is more or less triangular, and its length is about two times its width. The distance between the dorsal and occipital processes is 4–5.5 times the distance from the tip of the snout to the end of the occipital process. Genital papilla in males is elongated and pointed. This species grows to a maximum size of 47 cm and 1.2 kg (Luna, 1988f).

Nutritional Facts

Proximate Composition (% WW)

	TPN	Protein	Fat	Moisture	Ash
Male	2.66–4.01	16.67–25.03	0.27–0.78	78.61–80.12	1.05–1.53
Female	2.51–3.99	16.60–24.94	0.30–0.81	78.31–80.31	1.05–1.49

TPN, total protein nitrogen.
Source: Data from Bano (1977).

Clarias gariepinus Burchell 1822

Order: Siluriformes

Family: Clariidae

Common name: North African catfish.

Distribution: China, Thailand, Egypt, Uganda, Jordan, Lebanon, Israel, and Turkey, Africa, as well as in Europe, Asia, and South America (Pouomogne, 2014).

Habitat: Lakes, streams, rivers, swamps, and floodplains.

Description: Body of this species is elongated. Head is large, depressed, and bony with small eyes. Narrow and angular occipital process is seen. Gill openings are wide. Mouth is terminal and large. Four pairs of barbels are present. Dorsal and anal fins are long. Dorsal fin spine and adipose fin are absent. Anterior edge of pectoral spine is serrated. Caudal fin is rounded. Color varies from sandy-yellow through gray to olive with dark greenish brown markings, and belly is white. This species grows to a maximum size of 1.5 m and 60 kg (Pouomogne, 2014).

Nutritional Facts

Proximate Composition (% WW)

Protein	Lipid	Moisture	Ash
19.64	1.15	76.71	1.23

Fatty Acids (%)

C12:0 Lauric acid	3.1
C14:0 Myristic acid	4.2
C14:1 Myristoleic acid	0.2
C15:0 Pentadecanoic acid	0.3
C16:0 Palmitic acid	22.0
C16:1 Palmitoleic acid	3.6
C17:0 Heptadecanoic acid	0.7
C17:1 Heptadecenoic acid	0.2
C18:0 Stearic acid	8.1
C18:1 Oleic acid	26.0
C18:2 Linoleic acid	12.3
C18:3 Linolenic acid (omega-3)	1.0
C18:3 Linolenic acid (omega-6)	0.6
C18:4 Octadecatetraenoic acid	1.6
C20:0 Arachidic acid	0.2
C20:1 Gadoleic acid	2.5
C20:2 Eicosadienoic acid	0.6
C20:3 Eicosatrienoic acid (omega-3)	0.1
C20:3 Eicosatrienoic acid (omega-6)	0.6
C20:4 Arachidonic acid (omega-3)	0.3
C20:4 Arachidonic acid (omega-6)	0.6
C20:5 Eicosapentaenoic acid	1.0
C22:0 Behenic acid	0.1
C22:1 Cetoleic acid	1.3
C22:4 Docosatetraenoic acid	0.6
C22:5 Clupanodonic acid	1.0
C22:6 Docosahexaenoic acid	3.0

Source: Data from Osibona et al. (2006).

Minerals (% of Body Tissue)

P	Ca	K	Mg	Fe	Na	Cu	Zn
0.34	0.33	0.36	0.30	0.12	0.80	0.02	0.40

Source: Data from Fawole et al. (2007).

Major Amino Acids (mg/g)	
Glutamic acid	28.45
Aspartic acid	18.14
Lysine	17.00
Leucine	15.22
Arginine	10.89
Alanine	10.30
Medium Amino Acids (mg/g)	
Valine	8.53
Isoleucine	8.35
Glycine	8.10
Threonine	7.68
Serine	7.16
Phenylalanine	6.69
Proline	6.09
Methionine sulfone	5.06
Minor Amino Acids (mg/g)	
Histidine	4.37
Cystine	1.85
Tyrosine	1.85
Ornithine	1.03
Taurine	0.85
g-Aminobutyric acid	0.81
Hydroxyproline	0.45

Source: Data from Osibona (2011).

Heterobranchus bidorsalis Geoffroy Saint-Hilaire 1809

Order: Siluriformes

Family: Clariidae

Common name: Eel-like fattyfin cat-fish, African catfish.

Distribution: Africa: Nile, Chad Basin, Niger, Upper Volta, Guinea, Senegal, and Gambia (Geelhand, 2013b).

Habitat: Turbulent or fast-running streams.

Description: Head of this species is oval shaped to rectangular dorsally. Snout is broadly rounded. Eyes are rather laterally placed. Pectoral spine is smooth. Body and fins occasionally have irregularly placed spots. It can reach a length of 150 cm and weight of 30 kg (Geelhand, 2013b).

Nutritional Facts

Proximate Composition (% WW)

Moisture	Protein	Fat	CHO[a]	Ash
72.53	19.67	7.25	2.55	2.50

[a] Carbohydrate.

Gross Calorific Values (kcal/100 g)

Protein	Fat	CHO[a]	Total
15.85	36.31	1.17	53.55

[a] Carbohydrate.

Source: Data from Keremah and Amakiri (2013).

FAMILY: MOCHOKIDAE (SQUEAKERS OR UPSIDE-DOWN CATFISHES)

Hemisynodontis membranaceus Geoffroy Saint-Hilaire 1809 (= *Synodontis membranacea*)

Order: Siluriformes

Family: Mochokidae

Common name: Moustache catfish.

Distribution: Egypt, Nigeria, Gambia, Cameroon, Senegal, Chad, Malawi, Ghana, Niger, and Sudan (SeriouslyFish, 2014a).

Habitat: Deep shorelines of rivers and streams.

Description: This species can be recognized by the unusually wide membranes on its barbels that give the appearance of a moustache. This fish has a few large spots on its body that fade quickly with growth. There is a comb of very small spines near the point of the operculum, but they do not protrude through the skin. This fish may reach 50 cm size (Froese and Pauly, 2014d).

Nutritional Facts

Proximate Composition (% WW)

Moisture	Ash	Fat	Protein	Fiber
84.2	0.75	5.0	21.52	0.90

Minerals (mg/g)

Ca	K	Mg	P
2.89	0.71	0.29	0.027

Amino Acids (mg/g)

Alanine	6.61
Arginine	5.88
Aspartic acid	12.84
Glutamic acid	15.76
Glycine	6.25
Histidine	3.24
Isoleucine	4.65
Leucine	8.10
Lysine	10.62
Methionine	0
Phenylalanine	4.87
Proline	3.87
Serine	4.56

Source: Data from Effiong and Mohammed (2008).

Synodontis schall Bloch & Schneider 1801

Order: Siluriformes

Family: Mochokidae

Common name: Shield-head catfish.

Distribution: Northern Africa.

Habitat: Freshwater; deep, open water and in quite shallow water.

Description: This species has a shield on its body and has strong bony spines on the pectoral and dorsal fins. Its swim bladder is whitish in appearance. The roof of the front section of the swim bladder is directly attached to the backbone, and the muscles that are connected to it are able to compress and contract the swim bladder. This species is also distinguished by the length of its adipose fin, which is very long and runs from just past the base of the caudal fin to just before the caudal fin. This species grows to a maximum length of 49 cm (Araoye, 2000).

Nutritional Facts

Proximate Composition (g/100 g DM)

Lipid	Protein	Moisture[a]
17.3	59.8	73.0

[a] Wet weight.

Minerals (mg/g DM)

K	P	Ca	Na	Mg	Zn[a]	Fe[a]	Cu[a]
10,175	7370	3113	2805	696	53	60	1.4

[a] µg.

Essential Amino Acids (EAAs)
(µg/g DM)

Leucine	260.5
Isoleucine	141.6

Valine	158.3
Threonine	171.6
Lysine	300.3
Histidine	106.3
Methionine	124.6
Phenylalanine	156.8
Total EAAs	1420.0

Source: Data from Mohamed et al. (2010).

FAMILY: HETEROPNEUSTIDAE (AIRSAC CATFISHES)

Heteropneustes fossilis Bloch 1794

Order: Siluriformes

Family: Heteropneustidae

Common name: Asian stinging catfish.

Distribution: Asia: Pakistan and Sri Lanka to Myanmar (Susan, 1988a).

Habitat: Ponds, ditches, swamps, and marshes.

Description: Body of this species is elongated and compressed. Head is depressed and covered with an osseous plate at top and sides of the head. There are four pairs of barbels. Caudal fin is rounded. Body color is reddish brown or purplish brown. Matured specimens are, however, black in color. Using its swim bladder, it can breathe atmospheric air to a certain extent, allowing it to survive in some fairly hostile conditions. This species reaches a maximum size of 30 cm (SeriouslyFish, 2014b).

Nutritional Facts

Proximate Composition (% WW)

Moisture	Ash	Protein	Lipid
80.44	0.94	15.14	3.49

Source: Data from Ahmed et al. (2012).

Minerals (mg/100 g DM)

K	Na	Ca	P	Fe
229.9	40.3	65.0	621.3	2.3

Source: Data from Wimalasena and Jayasuriya (1996).

FAMILY: ICTALURIDAE (NORTH AMERICAN FRESHWATER CATFISHES)

Ictalurus punctatus Rafinesque 1818

Order: Siluriformes

Family: Ictaluridae

Common name: Channel catfish.

Distribution: Europe, Russian Federation, Cuba, and portions of Latin America (Stickney, 2014).

Habitat: Flowing waters; large streams with low or moderate current.

Description: This species has a cylindrical body with skin lacking in scales. Numerous small black spots are seen on the sides. Soft fin rays are seen with the exception of the dorsal and pectoral fins, which have spines. Adipose fin is present. Barbels are located below and at the corners of the mouth, with two barbels on the dorsal surface of the head anterior of the eyes and posterior of the snout. Upper jaw projects beyond the lower jaw. Deeply forked caudal fin and curved anal fin are present. Coloration is olive-brown to slate blue on the back and sides, shading to silvery-white on the belly. This species attains a maximum size of 132 cm and 26.3 kg (Stickney, 2014).

Nutritional Facts

Proximate Composition (g/100 g, WW)

Protein	Fat	Moisture	Ash	Energy[a]
16.3	5.4	77.3	1.05	118

[a] kcal.

Fatty Acids (% of Total Fatty Acids)

Saturated	23.76
Monoenes	43.75
Dienes	15.31
Trienes	4.07
n-3	3.74
n-6	21.46

Vitamins (µg/g)

Thiamin	1.9
Riboflavin	1.4
Pyridoxine	1.2
Folic acid	0.15
Niacin	13.5
Pantothenic acid	10.1
Choline	611
Ascorbic acid	2.5

Minerals (µg/g)

K	P	Na	Mg	Ca	Zn
3536	1799	404	224	91	5.9
Fe	Se	Cu	Mn	Co	
5.3	0.11	<0.3	<1.1	<1.1	

Source: Data from Robinson et al. (2001).

Amino Acids (g/100 g)

Arginine	6.67
Histidine	2.17
Isoleucine	4.29
Leucine	7.40

Continued

Lysine	8.51
Methionine	2.92
Cysteine	0.86
Phenylalanine	4.14
Tyrosine	3.28

Threonine	4.41
Tryptophan	0.78
Valine	5.15

Source: Data from Jarmolowicz and Zakês (2014).

FAMILY: SISORIDAE (SISORID CATFISHES)

Hemibagrus nemurus Valenciennes 1840 (= *Macrones nemurus*, *Mystus nemurus*)

Order: Siluriformes

Family: Sisoridae

Common name: Yellow catfish, Asian redtail catfish.

Distribution: Asia: Mekong, Chao Phraya, and Xe Bangfai Basins; Malay Peninsula, Sumatra, Java, Borneo (Pablico, 1997a).

Habitat: Rivers and streams, lakes, marshlands, and peat swamps; large muddy rivers with slow current and soft bottom.

Description: Head of this species is flattened rather than conical. Dorsal fin is depressed and does not reach the adipose fin. Pectoral fin is smooth in front, and its spines are serrated along the inner edge. Base of the adipose fin is shorter than that of the dorsal fin and about equal to that of the anal fin. Barbels are in four pairs. Body color is brown, often with a greenish sheen. Fins are gray with a violet tint. It grows to a maximum size of 65 cm (Pablico, 1997a).

Nutritional Facts

Proximate Composition (g/100 g Edible Portion, WW)

Energy[a]	Water	Protein	Fat	CHO[b]	Ash
118	76.2	16.6	5.5	0.6	1.1

[a] kcal.

[b] Carbohydrate.

Minerals (mg/100 g Edible Portion)

Ca	P	Fe	Na	K
11	192	0.5	38	253

Vitamins (µg per 100 g Edible Portion)

Retinol	Carotene	Thiamin	Riboflavin	Niacin	Vitamin C
47 µg	6 µg	0.07 mg	0.16 mg	1.6 mg	3.1 mg

Source: Data from Tee et al. (1989).

225

Amino Acids (mg/100 g)

Aspartic acid	4.98
Glutamic acid	5.11
Serine	3.79
Glycine	2.11
Histidine	3.34
Arginine	6.93
Threonine	5.46
Alanine	3.51
Proline	2.83

Tyrosine	2.41
Valine	5.89
Methionine	2.97
Cystine	1.28
Isoleucine	4.46
Leucine	6.34
Phenylalanine	4.47
Lysine	4.97

Source: Data from Hudu et al. (2010).

FAMILY: PANGASIDAE (SHARK CATFISHES)

Pangasianodon hypophthalmus Sauvage 1878

Order: Siluriformes

Family: Pangasiidae

Common name: Striped catfish.

Distribution: Asia: Mekong, Chao Phraya, Maeklong Basins (Sampang, 1999a).

Habitat: Riverine freshwater.

Description: Body of this species is long and latterly flattened with no scales. Head is relatively small. Mouth is broad with small sharp teeth on jaw, vomerine, and palatal bones. Eyes are relatively large. Two pairs of barbels are seen, and the upper is shorter than the lower. Fins are dark gray or black. There are six branched dorsal fin rays. Gill rakers are normally developed. Body is uniformly gray, but sometimes with a greenish tint and sides silvery. A dark stripe is noticed on the middle of the anal fin and in each caudal lobe. This species can reach a maximum size of 130 cm (Rainboth, 1996).

Nutritional Facts

Proximate Composition (% WW)

Moisture	Protein	Fat	Ash	CHO[a]
76.8	13.5	1.49	1.40	1.20

[a] Carbohydrate.

Source: Data from Babu et al. (2013).

Pangasius pangasius Hamilton 1822

226

Order: Siluriformes

Family: Pangasiidae

Common name: Pangas catfish.

Distribution: Asia: Indian subcontinent and Myanmar (Sampang, 1999b).

Habitat: Freshwater, brackish water; large rivers and estuaries.

Description: Body is elongated and laterally compressed. Snout is obtusely rounded. Upper jaw is longer than lower jaw and mouth gap is moderate. Two pairs of barbels are present. Dorsal spine is serrated anteriorly. Pectoral spine is comparatively stronger than dorsal spine and serrated internally. Caudal fin is deeply forked and lateral line is complete. Abdomen is silvery and yellowish green/dark on back. This species reaches a maximum size of 3 m. This species reaches a maximum size of 3 m (Sampang, 1999b).

Nutritional Facts

Proximate Composition (g/100 g WW)

	Moisture	Protein	Fat	Ash	CHO[a]	Energy[b]
Native pangas	56.11	26.06	14.79	1.11	1.80	244.67
Hybrid pangas	62.71	23.18	11.11	0.86	2.13	201.23

[a] Carbohydrate.
[a] kcal.

Minerals (mg/100 g)

	Native Pangas	Hybrid Pangas
Zn	0.88	0.75
Na	210.26	99.68
Ca	90.71	56.36
K	311.71	214.06
Fe	4.58	1.51
Cu	0.36	0.20
Mg	48.94	21.03
Al	0.71	0.89

Nonessential Amino Acids (g/100 g)

	Native Pangas	Hybrid Pangas
Aspartic acid	0.37	0.26
Serine	0.70	0.64
Glutamic acid	0.72	0.55
Glycine	3.12	2.24
Alanine	0.81	0.56
Tyrosine	0.42	0.28
Arginine	0.98	0.84

Source: Data from Monalisa et al. (2013).

Essential Amino Acids (g/100 g)

	Native Pangas	Hybrid Pangas
Lysine	3.24	2.61
Leucine	0.21	0.16
Isoleucine	1.10	0.81
Valine	0.17	0.09
Methionine	0.40	0.32
Threonine	0.86	0.77
Histidine	1.06	0.86

ORDER: CLUPEIFORMES (HERRINGS)

FAMILY: CLUPEIDAE (HERRINGS, SHADS)

Gudusia chapra Hamilton 1822

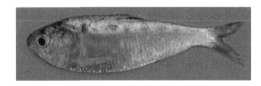

Order: Clupeiformes

Family: Clupeidae

Common name: Indian river shad.

Distribution: Asia: India, Bangladesh, Nepal, and Pakistan (Bailly, 1997d).

Habitat: Restricted to lowland, slow-flowing rivers, with sandy beds; ponds, beels, ditches, and inundated fields.

Description: Body is fairly deep and ventral profile is more convex than dorsal. Insertion of pelvic fins is just before dorsal fin origin. Upper jaw has a distinct median notch at center. Scales are small and 77–91 scales are seen in lateral series. A dark blotch is noticed behind the gill opening, often followed by a series of spots along the flank (Das et al., 2012).

Nutritional Facts

Proximate Composition (% WW)

Protein	Fat	Moisture	Ash
15.23	5.41	75.07	1.55

Source: Data from Mazumder et al. (2008).

Corica soborna Hamilton 1822

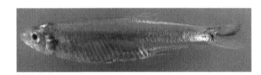

Order: Clupeiformes

Family: Clupeidae

Common name: Ganges river sprat.

Distribution: Asia: Indonesia, Brunei, Malaysia, and India (Bailly, 1997c).

Habitat: Rivers and estuaries and coastal areas.

Description: Body of the species is elongated and laterally compressed. Body is keeled with 11 + 7 scutes. There are 19–21 lower gill rakers. Last two rays of the anal fin form a separate finlet. There are 40–43 scales in lateral series. Caudal fin is forked and its lower lobe is slightly longer than the upper lobe. Body is brownish. A faint lateral band runs longitudinally. Caudal fin is with faint edges and a faint black spot is seen at its base (Chaki, 2013).

**Vitamin A and Minerals
(per 100 g Raw Cleaned Parts)**

Vitamin A (RAE)	Ca[a]	Fe[b]	Zn[b]
90	0.5	2.8	3.1

RAE, retinol activity equivalent.

[a] g.

[b] mg.

Source: Data from Thilsted (2012).

Tenualosa ilisha Hamilton 1822

Order: Clupeiformes

Family: Clupeidae

Common name: Hilsa shad.

Distribution: Indian Ocean: Persian Gulf eastward to Myanmar, including western and eastern coasts of India; Gulf of Tonkin, Vietnam, and southern Iran (Bailly, 1997m).

Habitat: Coastal waters and ascending rivers.

Description: This species has a moderately deep body that is laterally compressed with 30–33 scutes on the ventral side. Abdominal part is keeled with strong scutes. A distinct median notch is seen in the upper jaw. Gill rakers are fine and numerous, about 100–250 on the lower part of the arch. Fins are hyaline. A dark blotch is noticed behind the gill opening, followed by a series of small spots along the flank in juveniles. Color in live fish is silver shot with gold and purple. It grows to a maximum length of 60 cm and weight of 680 g (Bailly, 1997m).

Nutritional Facts

Proximate Composition (% WW)

	Moisture	Protein	Fat	Ash
G. chapra	76.01	16.78	4.55	1.70
C. soborna	77.91	17.31	3.59	1.68
T. ilisha	59.36	22.56	18.49	2.27

Source: Data from Begum and Minar (2012).

Limnothrissa miodon Boulenger 1906

Order: Clupeiformes

Family: Clupeidae

Common name: Lake Tanganyika sardine.

229

Distribution: Africa: Endemic to Lake Tanganyika (Bailly, 1997g).

Habitat: Lacustrine species, preferring open waters.

Description: Body of this species is fairly slender. Prepelvic scutes are not strongly keeled, beginning behind base of last pectoral fin ray. Lower gill rakers numbering 35–38 are long and slender. A distinct silver stripe is seen along the flank. Snout is broad with tapering sides, not concave when viewed from above. It grows to a maximum size of 17 cm. It has a large air bladder, which is responsible for its ability to move great vertical distances (Bailly, 1997g).

Nutritional Facts

Proximate Composition (g/100 g DM)

Protein	Lipid	CHO[a]	Moisture	Ash
67.22	6.75	5.03	5.44	15.56

[a] Carbohydrate.

Minerals (mg/g DM)

Ca	P	K	Na	Mg	Zn	Fe
306	175	45	26	19	1.23	0.38

Source: Data from Abdulkarim et al. (2013).

Stolothrissa tanganicae Regan 1917

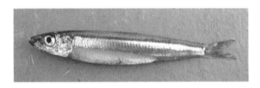

Order: Clupeiformes

Family: Clupeidae

Common name: Lake Tanganyika sprat.

Distribution: Africa: Endemic to Lake Tanganyika (Bailly, 1997l).

Habitat: Lacustrine species; Lake Tanganyika.

Description: Body of this species is slender. Prepelvic scute is not strongly keeled, beginning behind base of last pectoral fin ray. Maxilla blade is about 2.25 times as long as its shaft, but is not continued forward to the hind tip of the premaxilla; the second supra-maxilla is diamond shaped or more or less rhomboidal, approximately symmetrical. Lower gill rakers are long and slender. Scales are small, 36–46 in lateral series. A silver stripe is seen along flanks. Snout is narrow, appears concave when viewed from above. It grows to a maximum size of only 10 cm (Bailly, 1997l).

Nutritional Facts

Proximate Composition (g/100 g DM)

Protein	Lipid	CHO[a]	Moisture	Ash
63.31	8.64	7.41	6.76	13.88

[a] Carbohydrate.

Minerals (mg/g DM)

Ca	P	K	Na	Mg	Zn	Fe
294	169	46	26	17	0.98	0.21

Source: Data from Abdulkarim et al. (2013).

ORDER: ESOCIFORMES (PIKES AND PICKERELS)

FAMILY: ESOCIDAE (PIKES)

Esox lucius Linnaeus 1758

Order: Esociformes

Family: Esocidae

Common name: Northern pike.

Distribution: North America and Eurasia; Australia and New Zealand (Froese and Pauly, 2003a).

Habitat: Clear vegetated lakes, quiet pools, and backwaters of creeks and small to large rivers.

Description: Dorsal and anal fins are opposite and close to the tail fin. Pelvic fins are abdominal in position. Head is broad, although the snout is pointed. A series of large teeth is seen in the lower jaw, and dense but smaller teeth are present in the roof of the mouth. Color is usually greenish brown, flecked with lighter golden green to form curved lines and speckles on the back and sides. It is yellowish ventrally. It grows to a maximum size of 130 cm and 34 kg. The larger fish are females, and males rarely grow heavier than 6.3 kg (Froese and Pauly, 2003a).

Nutritional Facts

Proximate Composition (g/100 g WW)

Moisture	Protein	Fat	Ash	Cholesterol[a]
79.00	18.43	1.61	0.64	36.37

[a] mg/100 g.

Fatty Acids (% WW)

SFA	MUFA	PUFA	PUFA/SFA
39.90	31.66	28.15	0.71

Source: Data from Ljubojevic et al. (2013).

Protein Content of the Meat

% of Wet Weight	% of Dry Matter
18.4	90

Essential Amino Acid Composition (g/16 g N of the protein)

Arginine	7.4
Histidine	2.1
Isoleucine	5.3
Leucine	8.2
Lysine	9.2
Methionine	3.2
Phenylalanine	4.2
Threonine	5.0
Tryptophan	0.9
Valine	5.6

Lipid Content of the Meat

% of Wet Weight	% of Dry Matter
0.9	4

Major Elements in the Muscle (mg/100 g)

Na	K	Mg	Ca	P
75	304	28	20	215

Trace Elements in the
Muscle (µg/100 g)

Fe	Mn	Se
615	43	18

Vitamins in the Muscle
(µg/100 g)

A	B1	B2	B6
14	85	55	150

Source: Data from Steffens et al.
(2006).

ORDER: OSTEOGLOSSIFORMES (BONY TONGUES)

FAMILY: NOTOPTERIDAE (FEATHERBACKS OR KNIFEFISHES)

Notopterus notopterus Pallas 1769

Order: Osteoglossiformes

Family: Notopteridae

Common name: Bronze featherback.

Distribution: Pakistan, India, Nepal, Bangladesh, Myanmar, Laos, Cambodia, Thailand, Vietnam, Malaysia, and Indonesia (Froese and Pauly, 2014c).

Habitat: Fresh and brackish waters; lentic waters.

Description: Body of the species is highly compressed. Dorsal and ventral profiles are equally convex. Scales are minute. Dorsal fin is short. Anal fin is long and confluent with the caudal fin. Body color is silvery-white with many fine gray spots on the body and head. Lateral line is present with 230–240 scales. Juveniles possess a number of dark vertical bars on the body. It grows to a maximum length of 60 cm (Froese and Pauly, 2014c).

Nutritional Facts

Proximate Composition (% WW)

Moisture	Protein	Fat	Ash	CHO[a]
80.0	16.2	1.2	1.4	1.2

[a] Carbohydrate.

Amino Acids (g/16 g N)

Lysine	5.7
Histidine	1.7
Arginine	5.9
Aspartic acid	8.6
Threonine	3.9
Serine	3.6
Glutamic acid	11.3
Proline	3.9
Glycine	5.3
Alanine	5.4
Cystine	0.4
Valine	3.7
Methionine	2.5
Leucine	2.6
Isoleucine	3.1
Tyrosine	2.5
Phenylalanine	3.1

Source: Data from Zanariah and
Rehan (1988).

Chitala chitala Hamilton 1822

Order: Osteoglossiformes

Family: Notopteridae

Common name: Clown knifefish.

Distribution: Pakistan, India, Bangladesh, Nepal and Myanmar, Indonesia, Cambodia, Malaysia, and Thailand (Chaudhry, 2010).

Habitat: Rivers, lakes, beels, nullahs in the plains, reservoirs, canals, and ponds.

Description: This species has a long knife-like body. There are two nasal tentacles above the large, toothed mouth. Dorsal fin is flag-like and is usually located at the center of the body. The center ridge along the belly is distinctly serrated. Ventral fins are absent. It is overall silvery in color. It may have a series of golden or silvery bars along the back, resulting in a faint striped appearance. Additionally, it has a series of fairly small, sometimes indistinct, non-ocellated dark spots toward the far rear of the body (at the tail). This species reaches a maximum length of 122 cm (Weber and Beaufort, 1964).

Nutritional Facts

Proximate Composition (g/100 g Edible Portion, WW)

Energy[a]	Water	Protein	Fat	CHO[b]	Ash
105	75.7	20.1	2.5	0.5	1.2

[a] kcal.
[b] Carbohydrate.

Minerals (mg/100 g Edible Portion)

Ca	P	Fe	Na	K
87	241	0.2	20	260

Vitamins (per 100 g Edible Portion)

Retinol	Carotene	Thiamin	Riboflavin	Niacin	Vitamin C
92 μg	0 μg	0 mg	0.03 mg	3.9 mg	5.2 mg

Source: Data from Tee et al. (1989).

FAMILY: ARAPAIMIDAE (BONY TONGUES)

Heterotis niloticus Cuvier 1829

Order: Osteoglossiformes

Family: Arapaimidae

Common name: African arowana.

Distribution: Throughout Africa (Sahelo-Sudanese region, Senegal, and Gambia, as well as parts of eastern Africa) (Froese and Pauly, 2014b).

Habitat: Open water of rivers and lakes.

Description: It is a long-bodied fish with short head, large scales, long dorsal and anal fins set far back on the body, and a rounded caudal fin. It has an elongated and robust body with a height 3.5–5 times the standard length (SL). Lips are thick, and there is a dermal flap on the border of the gill cover. Adults are gray, brown, or bronze in color, but juveniles often have dark longitudinal bands. It grows to a maximum length of 1 m and weight of 10.2 kg (Froese and Pauly, 2014b).

Nutritional Facts

Proximate Composition (% DM)

Protein	Fiber	Ash	Moisture
46.28	0.12	9.00	7.90

Minerals (% DM)

P	Ca	K	Mg	Fe	Na	Cu	Zn
0.38	0.40	0.32	0.29	0.14	0.74	0.03	0.30

Source: Data from Fawole et al. (2007).

FAMILY: MORMYRIDAE (ELEPHANT FISHES)

Gnathonemus tamandua Günther 1862 (= *Campylomormyrus tamandua*)

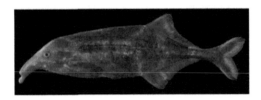

Order: Osteoglossiformes

Family: Mormyridae

Common name: Trunkfish, blunt-jawed elephantnose.

Distribution: Western Africa; Niger and Zaire (Mongabay.com, 1994–2013).

Habitat: Murky water with submerged wood.

Description: It is an elongated, laterally compressed species with a long elephant-like snout, and the mouth is located near the end of this proboscis. Lower part of the proboscis protrudes past the mouth. Dorsal and ventral profiles are symmetric, with the anal and dorsal fins located across from one another. Caudal fin is forked. Coloration of the body is brown-gray with several brownish white markings. This species reaches a maximum size of 40 cm (Günther, 1864; Mongabay.com, 1994–2013).

Nutritional Facts

Proximate Composition (% DM)

Moisture	Ash	Protein	Fat	EE	NFE
10.51	25.50	10.37	16.56	35.07	1.60

EE, ether extract; NFE, nitrogen-free extract.
Source: Data from Olele (2012).

234

Mormyrus rume Valenciennes 1847

Order: Osteoglossiformes

Family: Mormyridae

Common name: Mormyrid.

Distribution: Africa: Gambia, Senegal, Niger, Volta, and Chad Basins (Froese and Pauly, 2011b).

Habitat: Rivers.

Description: Depth of body of this species is 3½–5¼ times in total length, and length of head is 4⅓–5 times. Upper profile of head is descending in a straight line or feeble curve. Snout is as long as post-ocular part of head. Mouth is very small, with thick lips. Teeth are notched or truncate, with 5 or 7 in upper jaw and 6–10 in lower; eye is moderate, with a diameter 1⅔–2 times interorbital width. Dorsal fin originates considerably in advance of the base of ventrals. Anal fin originates at a nearly equal distance from the base of the pectoral and the base of the caudal. Pectoral fin is obtusely pointed. Caudal fin is densely scaled, with obtusely pointed lobes. Body is olive or purplish brown above and white beneath. It grows to a maximum size of 1 m and 5.3 kg (Boulenger, 1899).

Nutritional Facts

Proximate Composition (% DM)

Protein	Fiber	Ash	Moisture	Energy[a]
38.6	0.34	10.0	7.50	349.49

[a] kcal/100 g.

Minerals (mg/g)

Fe	Cu	Ni	Na	K	Zn	Pb
0.14	0.036	0.036	3.6	9.4	0.080	0.018

Source: Data from Onyia et al. (2010).

ORDER: GADIFORMES (CODS, HAKES, GRENADIERS, MORAS)

FAMILY: LOTIDAE (HAKES, BURBOTS)

Lota lota Linnaeus 1758

Order: Gadiformes

Family: Lotidae

Common name: Burbot.

Distribution: Europe (Luna, 1988l).

Habitat: Flowing waters and large, deep lakes, as well as large rivers with slow-moving current; estuaries of large lowland rivers and small mountain streams.

Description: In this species, its long second dorsal fin is at least six times as long as the first, and a single barbel is seen on the chin. Gill rakers are short. First dorsal is short and second dorsal and anal fins are joined to the caudal. Pectorals are short and rounded and caudal is rounded. Color is yellow, light tan to brown, with a pattern of dark brown or black on the body, head, and fins. Pelvic fins are pale and others are dark and mottled. This species grows to a maximum size of 152 cm (Luna, 1988l).

Nutritional Facts

Proximate Composition (% WW)

Moisture	Ash	Protein	Fat
64.5	1.4	16.2	9.4

Source: Data from Sarma et al. (2011).

ORDER: SALMONIFORMES (SALMON)

FAMILY: SALMONIDAE (SALMON)

Oncorhynchus tshawytscha Walbaum 1792

Order: Salmoniformes

Family: Salmonidae

Common name: Chinook salmon.

Distribution: United States; Siberia and south to Hokkaido Island, Japan (NOAA, 2014).

Habitat: Freshwater, marine water; diadromous; streams and rivers.

Description: Chinook salmon is the largest of any salmon with a maximum length of 150 cm and weight of 59 kg. This species is very similar to coho salmon in appearance (except for the large size). While at sea, it has blue-green back and silver flanks. Small black spots are seen on both lobes of the tail, and black pigment along the base of the teeth. Adults migrate from a marine environment into the freshwater streams and rivers of their birth in order to mate (called anadromy). They spawn only once and then die (semelparity) (NOAA, 2014).

Oncorhynchus gorbuscha Walbaum 1792

Order: Salmoniformes

Family: Salmonidae

Common name: Pink salmon.

Distribution: Arctic and Pacific drainages from Mackenzie River delta, Northwest Territories, Canada, to Sacramento River drainage, in California; occasionally as far as La Jolla, southern California; also in Northeast Asia. On Asia side, from North Korea to Jana and Lena drainages in Artic Russia (Torres, 1991i).

Habitat: Freshwater, brackish water, and marine water; diadromous; streams and rivers.

Description: This species is distinguished by the presence of large black spots on the back and on both lobes of the caudal fin. Body is fusiform, streamlined, and somewhat laterally compressed. It is, however, moderately deeper in breeding males. Mouth is large, terminal, and directed upwards and forward. Upper jaw reaches the posterior edge of orbit. Snout is rounded and narrow. Lips are fleshy and teeth are small and weak. However, in breeding males mouth becomes enlarged with well-developed teeth. Further, upper jaw in these males becomes hooked downwards. Adipose fin is large. Pelvic fins have axillary process. Fish in the sea are steel blue to blue-green on the back, silver on the sides, and white on the belly. Large oval spots are present on the back, adipose fin, and both lobes of the caudal fin. It grows to a maximum length of 76 cm and weight of 6.8 kg (Torres, 1991i).

Nutritional Facts

Proximate Composition (per 100 g edible portion, WW)

	O. tshawytscha	*O. gorbuscha*
Water (g)	71.64	75.52
Energy (kcal)	179	127
Protein (g)	19.93	21.31
Total lipid (g)	10.43	4.40

Amino Acids (g/100 g)

	O. tshawytscha	*O. gorbuscha*
Tryptophan	0.225	0.221
Threonine	0.879	1.066
Isoleucine	0.924	0.954
Leucine	1.630	1.562
Lysine	1.842	1.759
Methionine	0.594	0.577
Cystine	0.215	0.159
Phenylalanine	0.783	0.845
Tyrosine	0.677	0.742
Valine	1.033	1.100
Arginine	1.200	1.287
Histidine	0.590	0.543
Alanine	1.213	1.308
Aspartic acid	2.054	2.575
Glutamic acid	2.994	2.903
Glycine	0.963	1.263
Proline	0.709	0.867
Serine	0.818	0.905
Total amino acids	19.34	20.64

Lipids (per 100 g)

	O. tshawytscha	*O. gorbuscha*
Saturated fatty acids (g)	3.100	0.810
Monounsaturated fatty acids (g)	4.399	1.348
Polyunsaturated fatty acids (g)	2.799	0.811
20:5 n-3 (EPA) (mg)	385	233
22:6 n-3 (DHA) (mg)	944	333
Cholesterol (mg)	50	74

Minerals (mg/100 g)

	O. tshawytscha	*O. gorbuscha*
Calcium (Ca)	26.00	7.00
Iron (Fe)	0.25	0.38
Magnesium (Mg)	95	22
Phosphorus (P)	289	261
Potassium (K)	394	366
Sodium (Na)	47	50
Zinc (Zn)	0.44	0.39
Copper (Cu)	0.04	0.06
Manganese (Mn)	0.02	0.01
Selenium (Se)	36.5[a]	31.4[a]

[a] µg.

Vitamins (mg/100 g)

	O. tshawytscha	*O. gorbuscha*
Vitamin C	4.0	0.0
Thiamin	0.05	0.08
Riboflavin	0.11	0.11
Niacin	8.42	8.00
Pantothenic acid	0.75	1.03
Vitamin B6	0.40	0.61
Folate	30[a]	4[a]
Choline	ND	94.6
Vitamin B12	1.30	4.15
Vitamin A	453[b]	117[b]
Vitamin E	1.22	0.40
Vitamin D	ND	435[b]
Vitamin K	ND	0.40[a]

ND, no data.
[a] µg.
[b] IU.
Source: Data from Tacon and Metian (2013).

Oncorhynchus keta Walbaum 1792

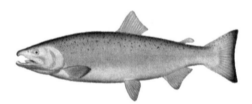

Order: Salmoniformes

Family: Salmonidae

Common name: Chum salmon.

Distribution: North Pacific: Korea, Japan, Okhotsk, and Bering Sea; Arctic Alaska south to San Diego, California; Asia: Iran (Pablico, 1997b).

Habitat: Ocean and coastal streams.

Description: This species is distinguished from other related species by the lack of distinct black spots on the back and tail and by the presence of 18–28 short, stout, smooth gill rakers on the first arch. Caudal fin is truncate to slightly emarginate. Large individuals are steel blue dorsally, with speckles of black. Sides are silvery and ventral is silvery to white. Males have tinges of black on the tips of their caudal, anal, and pectoral fins. This species grows to a maximum size of 1 m and 15.9 kg (Pablico, 1997b).

Nutritional Facts

Proximate Composition (% WW)

Moisture	Ash	Protein	Fat
74.1	1.18	21.3	3.86

Source: Data from Sarma et al. (2011).

Oncorhynchus kisutch Walbaum 1792

Order: Salmoniformes

Family: Salmonidae

Common name: Coho salmon.

Distribution: North Pacific: Russia, Japan, Alaska, and Bay in Baja California, Mexico (Pablico, 1997c).

Habitat: First half of their life cycle is in streams and small freshwater tributaries; spawning in small streams; remainder of the life cycle is in estuarine and marine waters.

Description: This species is characterized by the presence of small black spots on the back and on the upper lobe of the caudal fin, and by the lack of dark pigment along the gum line of the lower jaw. Gill rakers are rough and widely spaced. Lateral line is nearly straight. Adipose fin is slender. Individuals in the sea are dark metallic blue or greenish on the back and upper sides, a brilliant silver color on middle and lower sides, and white below. Small black spots are present on the back and upper sides and on the upper lobe of the caudal fin. This species grows to a maximum size of 108 cm and 15.2 kg (Pablico, 1997c).

Nutritional Facts

Proximate Composition (g/100 g WW)

Moisture	Protein	Lipid	Ash
66.9–77.2	20.0–22.8	1.6–2.5	1.6–2.5

Minerals (mg/100 g)

P	Ca	Fe
283–361	10–24	2–6

Fatty Acids (g/100 g)

C18:1	C22:6	C20:5	C16:0	C20:1
18.6	13.8	12.0	10.2	8.4

Vitamins (mg/100 g)

Niacin	7.34
Vitamin A	24.39[a]
Thiamine	0.08–0.10
Riboflavin	0.12–0.14
Ascorbic acid	0.8–1.4
Folic acid	0.01–0.02

[a] Retinol equivalents.

Source: Data from Vinagre et al. (2011).

Oncorhynchus mykiss **Walbaum 1792 (= *Salmo gairdneri*)**

Order: Salmoniformes

Family: Salmonidae

Common name: Rainbow trout.

Distribution: Native to Alaska, Baja California, and Mexico; Canada, United States (Torres, 1991j).

Habitat: Lakes, rivers, and streams.

Description: Body is elongated and somewhat compressed, especially in larger fish. No nuptial tubercles are seen. Stream residents and spawners are darker in color. Lake residents are lighter, brighter, and more silvery. This species grows to a size of 120 cm and 25.4 kg (Torres, 1991j).

Nutritional Facts

Proximate Composition (% WW)

Moisture	Protein	Lipid	Ash	Cholesterol[a]
71.65	19.60	4.43	1.36	35.04

[a] mg/100 g.

Fatty Acids (% of Total Fatty Acids)

SFA	27.65
MUFA	35.56
PUFA	23.09
n-3	15.64
n-6	7.45

Source: Data from Celik et al. (2008).

Major Elements in the Muscle (mg/100 g)

Na	K	Mg	Ca	P
63	413	26	12	245

Trace Elements in the Muscle (µg/100 g)

Fe	Mn	Se
441	30	25

Vitamins in the Muscle (µg/100 g)

A	B1	B2	B6
32	84	76	0

Source: Data from Steffens et al. (2006).

Amino Acids (g/100 g)

Arginine	6.41
Histidine	2.96
Isoleucine	4.34
Leucine	7.59
Lysine	8.49
Methionine	2.88
Cysteine	0.80
Phenylalanine	4.38
Tyrosine	3.38
Threonine	4.76
Tryptophan	0.93
Valine	5.09

Source: Data from Jarmolowicz and Zakês (2014).

Oncorhynchus nerka Walbaum 1792

Order: Salmoniformes

Family: Salmonidae

Common name: Sockeye salmon.

Distribution: North Pacific: Northern Japan to Bering Sea and to Los Angeles, California; Alaska, Yukon Territory, and British Columbia in Canada, and Washington and Oregon in United States (Kesner-Reyes, 2002a).

Habitat: Lakes, streams.

Description: This species is distinguished by the long, fine, serrated, closely spaced gill rakers on the first arch and by its lack of definite spot on the back and tail. Body is fusiform, streamlined, and laterally compressed. Body depth is moderate, slightly deeper in breeding males. Head is bluntly pointed and conical.

Eyes are rather small, and their position is variable with sex and condition. Snout is somewhat pointed. Lateral line is straight. Caudal fin is emarginated. Prespawning fish are dark steel blue to greenish blue on the head and back, silvery on the sides, and white to silvery on the belly. This species grows to a maximum size of 84 cm and 7.7 vkg (Kesner-Reyes, 2002a).

Nutritional Facts

Proximate Composition (% WW) and Energy (kJ/g)

	Moisture	Fat	Protein	Energy
Muscle of female	63.9	14.8	19.9	9.4
Muscle of male	63.4	15.4	19.8	9.6

Source: Data from Hendry (2000).

Salmo salar Linnaeus 1758

Order: Salmoniformes

Family: Salmonidae

Common name: Atlantic salmon.

Distribution: North Atlantic Ocean: temperate and arctic zones in northern hemisphere; western Atlantic (ranges from western Greenland and the coastal drainages of Quebec, Canada, to Connecticut) (Kesner-Reyes, 2002b).

Habitat: Streams, rivers.

Description: Body of this species is elongated, but becomes deeper with age.

Caudal peduncle is slender. Tip of upper jaw reaches to hind margin of eye, but not beyond. Jaws in adult males become greatly hooked just before and during breeding. A staggered line of teeth is seen on the shaft of the vomer, but none on the head of the vomer. Caudal fin is fairly deeply forked. Scales are small. Color of body is brown or green-blue at back. Flanks are silvery and belly is white back. Flanks above the lateral line (rarely below it) have X-shaped black spots. In freshwater, flanks are greenish or brown, mottled with red or orange, and with large dark spots with lighter edges. This species grows to a maximum size of 150 cm and 39 kg (Jones, 2014).

Nutritional Facts

Proximate Composition (% WW)

Moisture	Ash	Protein	Fat
75–82	NA	13–17	NA

NA, not available.

Source: Sarma et al. (2011).

Salmo trutta Linnaeus 1758

Order: Salmoniformes

Family: Salmonidae

Common name: Brown trout, sea trout.

Distribution: Throughout Europe; North Sea and the Baltic Sea (Fishbase 2003); Italy, Romania, France, the United Kingdom, Germany, Denmark, and Bosnia and Herzegovina (Vandeputte and Labbé, 2012).

Habitat: Cold streams, rivers, and lakes.

Description: This species has a fusiform body shape. Its adipose fin is with red margin. Highly variable color—from blue-gray to yellowish brown for resident river trout and silvery for lake and sea morphs, with black spots and red spots mostly in riverine forms. The distinguishing feature of this species is its upper maxilla, which extends mostly after the eye. Further, it has spots under the lateral line. This species grows to a maximum size of 50 cm and 2 kg (Vandeputte and Labbé, 2012).

Nutritional Facts

Proximate Composition (% WW)

Protein	Lipid	Moisture	Ash
18.1	2.7	74.8	1.6

Fatty Acids (%)

SFA	MUFA	PUFA	n-6	n-3	n-3/n6
27.7	28.6	26.4	5.9	20.5	3.5

Source: Data from Yeşilayer and Genç (2013).

Salmo macrostigma Duméril 1858
(= *Salmo trutta macrostigma*)

Order: Salmoniformes

Family: Salmonidae

Common name: Salmon.

Distribution: Africa: Restricted to Algeria (Geelhand, 2013d).

Habitat: Marine, freshwater, and brackish water.

Description: Dorsal fin of this species is slightly higher than long and is located more posteriorly than in other *Salmo* species. Caudal fin is more forked. There are 8–12 round, black spots (parr marks) on the sides, and such spots are more prominent in juveniles and are retained in adults. It grows to a maximum size of 60 cm (Geelhand, 2013d).

Nutritional Facts

Proximate Composition (% WW)

Moisture	Protein	Lipid	Ash
77.82	18.24	2.50	1.44

Free Amino Acids (FAAs) (mg/g WW)

Aspartic acid	0.10
Glutamic acid	0.82
Glycine	4.90
Proline	1.13
Alanine	0.23
Tyrosine	0.56
Arginine	0.06
Lysine	0.80
Methionine	1.43
Valine + isoleucine	0.38
Leucine	0.55
Phenylalanine	0.23
Nonessential FAAs	7.75
Essential FAAs	3.46
Total FAAs	11.21

Source: Gunlu and Gunlu (2014).

Fatty Acids (% of Total Fatty Acids)

SFA	32.26
MUFA	33.01
Omega-3 PUFA	29.12
Omega-6 PUFA	5.60
PUFA	34.72
Omega-3/omega-6 ratio	5.20
EPA + DHA	17.81

Source: Data from Ateş et al. (2013).

Salvelinus fontinalis Mitchill 1814

Order: Salmoniformes

Family: Salmonidae

Common name: Brook trout.

Distribution: North America: Most of eastern Canada from Newfoundland to western side of Hudson Bay; south in Atlantic, Great Lakes, and Mississippi River Basins to Minnesota and northern Georgia in the United States. South America: Argentina (Torres, 1991n).

Habitat: Marine, freshwater, and brackish water.

Description: This species is distinguished by the combination of dark green marbling on its back and dorsal fin and by the red spots with blue halos on its sides. Caudal fin is nearly straight or with a shallow indentation. Color of the body varies, but generally rather green to brownish on back, marked with paler vermiculations or marbling that extends onto the dorsal fin and sometimes the caudal. Sides are lighter than back, marked with numerous pale spots and some red spots, each of the latter surrounded by a blue halo. Anal, pelvic, and pectoral fins have a white leading edge followed by a dark stripe, and the rest of the fins are reddish. This species grows to a maximum size of 86 cm (Torres, 1991n).

Nutritional Facts

Proximate Composition (% WW)

Moisture	Ash	Protein	Fat
74.3	1.3	21.5	3.4

Source: Data from Sarma et al. (2011).

Salvelinus namaycush Walbaum 1792

Order: Salmoniformes

Family: Salmonidae

Common name: Lake trout.

Distribution: Northern North America; principally Canada, but also Alaska and, to some extent, the northeastern United States.

Habitat: Lakes.

Description: Two dorsal fins including one adipose fin are present in this species. Light spots are seen on darker gray background, and lower fins are edged with white. Tail is forked. It grows to a maximum size of 127 cm and 46.3 kg.

Nutritional Facts

Proximate Composition and Phosphorus (% WW)

Moisture	Protein	Lipid	Ash	P
80.33	14.27	3.62	2.09	0.41

Source: Data from Gunther et al. (2005).

Coregonus albula Linnaeus 1758

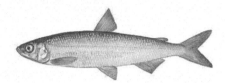

Order: Salmoniformes

Family: Salmonidae

Common name: White fish.

Distribution: The native range of this species is within drainages connected to the North and Baltic Seas, between the British Isles in the west and the Petchora drainage (Russia) in the east; some populations are also found in drainages to the White Sea and in lakes of the upper Volga drainage (www.cabi.org).

Habitat: Freshwater, brackish water, and marine water; deeper lakes, stillwaters.

Description: It is a small, streamlined, and slim fish with a bluish green back, a white belly, and silvery flanks. Fins are gray in color, becoming darker toward margins. It has large eyes, a relatively small mouth, and an adipose fin. It grows to a maximum size of 48 cm and 1 kg (Casal, 1995a; ARKive).

Nutritional Facts

Essential Amino Acid Composition (g/16 g N) of the Protein

Arginine	5.6
Histidine	1.9
Isoleucine	5.3
Leucine	6.9
Lysine	8.9
Methionine	3.1
Phenylalanine	3.5
Threonine	5.0
Tryptophan	1.0
Valine	5.9

n-3 and n-6 Content of Total Fatty Acids (%)

Σn-3	Σn-6	n-3/n-6
32.4	10.5	3.1

Source: Data from Steffens et al. (2006).

ORDER: CHARACIFORMES (CHARACINS)

FAMILY: CHARACIDAE (CHARACINS)

Brycon orbignyanus Valenciennes 1850

Phylum: Chordata

Class: Actinopterygii

Order: Characiformes

Family: Characidae

245

Common name: Piracanjuba.

Distribution: South America: La Plata Basin (Bailly, 1997b).

Habitat: Rivers of large and medium size, as well as small lakes connected to them.

Description: Females of this species may grow to a maximum length of 2 m and weight of 10 kg, and males grow to 60 cm and 3.5 kg, respectively. Body is elongated, and its dorsal part is slightly higher in older specimens. Color of the body is blue-green, while the fins are orange and bright colored. Caudal fin is black. It is omnivorous, feeding on fruits, seeds, and insects, and small fish.

Nutritional Facts

Proximate Composition (% DM) (Grown at 32% DW Protein and 12.1 DW[a] Lipid Diets)

Moisture	Protein	Lipid	Ash	Fiber	Energy[b]
23.02	32.65	12.36	4.33	11.49	21.25

[a] Dry weight.

[b] kJ/g.

Source: Data from Borba et al. (2003).

Piaractus brachypomus G. Cuvier 1818

Order: Characiformes

Family: Characidae

Common name: Red-bellied pacu.

Distribution: Tropical America; Orinoco and Amazon River Basins, South America (USGS, 2013).

Habitat: Marginal lagoons and ponds; upstreams.

Description: This species possesses a unique rare feature, viz., the adipose fin is rayed in adults. Body of juveniles is spotted and the fins are dark edged. They also have a distinctive large blotch on the opercle. Adults are uniformly dark or marbled. It grows to a maximum length of 88 cm and weight of 25 kg (U.S. Fish and Wildlife Service, 2012; Environmental Institute of Houston Invasive Species Inventory).

Nutritional Facts

Proximate Composition (% DM)

Moisture[a]	Protein	Fat	Ash
73.91	66.49	6.96	20.58

[a] Fresh weight.

Minerals (%) and Energy (kcal/kg)

Ca	P	Energy
13.87	3.07	2884

Source: Data from Barua and Chakraborty (2011).

FAMILY: CITHARINIDAE (LUTEFISHES)

Citharinus cithrus Geoffroy St. Hilaire 1809

Order: Characiformes

Family: Citharinidae

Common name: Moonfish.

Distribution: Africa: Gambia, Senegal, Niger, Volta, Ouémé and Chad Basins (Daget, 1984).

Habitat: Lakes, large rivers.

Description: Body shape is short and deep. Dorsal head profile is clearly concave. Mouth is terminal. Eyes are protected by small adipose eyelids. Body is silvery. Pectoral and superior lobes of the caudal fin are grayish. Inferior lobe of the caudal and anal fins are red. Basal part of adipose is blackish (Gosse and Paugy, 2003).

Nutritional Facts

Proximate Composition (% DM)

Protein	Fiber	Ash	Moisture
21.62	0.75	1.35	17.16

Minerals (mg/g)

Ca	K	Mg	P	Zn	Fe	Na
2.77	0.63	0.21	0.023	0.042	0.14	2.9

Source: Data from Effiong and Fakunle (2011).

Amino Acids (mg/g)

Alanine	4.06
Arginine	3.89
Aspartic acid	10.0
Glutamic acid	15.6
Glycine	2.52
Histidine	3.07
Isoleucine	2.49
Leucine	0
Lysine	8.39
Methionine	3.04
Phenylalanine	2.54
Proline	2.76
Serine	2.57

Source: Data from Effiong and Mohammed (2008).

Brycinus nurse Rüppell 1832 (= *Alestes nurse*)

247

Order: Characiformes

Family: Citharinidae

Common name: Nurse tetra.

Distribution: Widely distributed in West Africa (Geelhand, 2013a).

Habitat: Rivers, lakes, irrigation canals, and fringing vegetation.

Description: In this species, dorsal fin originates at about the same level as the pelvic fin insertion. Sexual dimorphism affects anal fin shape. There are 24–34 lateral line scales and 14–20 gill rakers on the lower limb of the first gill arch. A total of only eight teeth are present in the outer premaxillary row. Snout is short and is more than three times the head length. There are 10–11.5 predorsal scales and flanks are without lateral band. It grows to a maximum length of 25 cm and weight is 200 g (Geelhand, 2013a).

Nutritonial Facts

Proximate Composition (% WW)

Moisture	Protein	Fiber	CHO[a]	Ash	Energy[b]
72.26	20.26	0.93	0.67	2.47	0.625

[a] Carbohydrate.
[b] mg/kg DM.
Source: Data from Anene et al. (2013).

ORDER: ELOPIFORMES (TARPONS, TENPOUNDERS)

FAMILY: MEGALOPIDAE (OXEYE HERRINGS, TARPONS)

Megalops atlanticus Valenciennes 1847
(= *Tarpon atlanticus*)

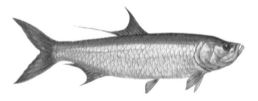

Order: Elopiformes

Family: Megalopidae

Common name: Tarpon.

Distribution: Eastern Atlantic: Senegal to Angola, with exceptional occurrences in Portugal, Azores, and Atlantic coast of southern France. Western Atlantic: North Carolina to Bahia, Brazil (Luna, 1988m).

Habitat: Coastal waters, bays, estuaries, and mangrove-lined lagoons; enter river mouths and bays and travel upstream into freshwater.

Description: In this species its vertical, silvery sides are made up of large scales. It has a superior mouth, with the lower mandible extending far beyond the gape. Fins are without spines, but they are all made of soft rays. Dorsal fin has 13–15 soft rays, with the last ray greatly elongated into a heavy filament. Caudal is deeply forked, and its lobes

248

appear equal in length. Anterior portion of the anal fin is deep and triangular. It has large pelvic fins and long pectoral fins. Sides and belly are predominantly silvery. Dorsally, it is dark blue to greenish black. However, the color may appear brownish or brassy for individuals inhabiting inland waters. Dorsal and caudal fins have dusky margins and often appear dark.

It grows to a maximum size of 250 cm and 161 kg (Morey, 2014; Luna, 1988m).

Nutritional Facts

Proximate Composition (% WW)

Moisture	Ash	Protein	Fiber	Fat	CHO[a]
35.95	14.36	25.15	0.65	0.88	23.01

[a] Carbohydrate.

Source: Data from Emmanuel et al. (2011).

ORDER: TETRAODONTIFORMES (PUFFERFISHES AND ALLIES)

FAMILY: TETRAODONTIDAE (PUFFERFISHES, TOADFISHES)

Tetraodon cutcutia Hamilton 1822

Order: Tetraodontiformes

Family: Tetraodontidae

Common name: Ocellated pufferfish.

Distribution: Asia: India, Bangladesh, Sri Lanka, Myanmar, and Malay Archipelago (Bailly, 1997n).

Habitat: Freshwater, brackish water; ponds, beels, canals, and rivers.

Distribution: This species has a broad head and back tapers abruptly to tail. Mouth opening is a little inferior with two large teeth in each jaw. Gill opening is very much reduced and is restricted in front of the pectoral base.

Eyes are large. Dorsal fin is placed well back and above the origin of the anal fin. All fins are rounded. Body is greenish yellow with white abdomen. A light band is seen between the eyes. A large black ocellus surrounded by a light edge is noticed on the side anterior to the origin of the anal fin. It grows to a maximum length of only 15 cm (Bashar, 2010c).

Proximate Composition (g/100 g WW) (Dorsal Muscle)

Moisture	Ash	Oil	Protein
30.94	9.1	35.6	23.39

249

Polysaccharides and Sugar (g/100 g WW)

Polysaccharides	Sugar
0.051	0.072

Minerals (mg/100 g WW)

Ca	P	Fe
40.5	52.5	46.5

Source: Data from Uddin (2012).

Tetraodon lineatus Linnaeus 1758

Order: Tetraodontiformes

Family: Tetraodontidae

Common name: Nile puffer, striped puffer, band puffer, globe fish.

Distribution: Nile, Chad Basin, Niger, Volta, Gambia, Geba and Senegal in Africa (Animal-world).

Habitat: Large rivers, open water, weed beds, and vegetated fringes in water.

Description: It is a stocky elongated fish covered with short prickles and has bright orange-red eyes. Body is brownish gray on the back, gradually becoming lighter toward the under parts, ending with a whitish belly. There are a series of light, often golden colored, horizontal stripes running from the pectoral fins back across the tail. These fishes are known to change color depending on mood. This species grows to a maximum length of 45 cm (Animal-world).

Nutritional Facts

Proximate Composition (g/100 g DM)

Lipid	Protein	Moisture[a]
1.8	79.1	80.0

[a] Wet weight.

Essential Amino Acids (EAAs) (μg/g DM)

Leucine	177.6
Isoleucine	97.0
Valine	110.4
Threonine	114.5
Lysine	202.3
Histidine	78.9
Methionine	88.3
Phenylalanine	91.6
Total EAAs	960.5

Minerals (per g DM)

K (mg)	P (mg)	Ca (mg)	Na (mg)	Mg (mg)	Zn (μg)	Fe (μg)	Cu (μg)
9990	7885	5880	2035	751	88	61	1.3

Source: Data from Mohamed et al. (2010).

ORDER: SYNBRANCHIFORMES (SWAMP EELS)

FAMILY: MASTACEMBELIDAE (SPINY EELS)

Macrognathus pancalus Hamilton 1822
(= *Mastacembelus punctatus*)

Order: Synbranchiformes

Family: Mastacembelidae

Common name: Striped spiny eel, barred spiny eel, zigzag eel, yellow fin spiny eel, Indian spiny eel.

Distribution: Asia: Pakistan, India, and Bangladesh; reported from Nepal (Torres, 1991g).

Habitat: Slow and shallow waters of rivers of plains and estuaries; canals, streams, beels, ponds, and inundated fields.

Description: Males of this species are more slender and often smaller than the females. The terminal rim of the anterior nasal tube consists of six flaps of skin. The overall darkness of the fish and the actual amount of yellow in the coloration is variable (Fishbase, 2005). It is a carnivore feeding on live foods such as earthworms, black worms, krill, and plankton. It grows to a maximum size of only 18 cm (Britz, 1996).

Nutritional Facts

Proximate Composition (% WW)

Moisture	Ash	Protein	Lipid
76.09	1.62	18.03	4.25

Source: Data from Ahmed et al. (2012).

Proximate Composition (% DM)

Moisture	CHO[a]	Protein	Fat	Ash
2.94	21.11	18.25	14.30	1.20

[a] Carbohydrate.

Amino Acids (% DM)

Methionine	Tryptophan
1.22	1.02

Source: Data from Ali et al. (2013).

Mastacembelus mastacembelus Banks & Solander 1794

Order: Synbranchiformes

Family: Mastacembelidae

Common name: Mesopotamian spiny eel.

251

Distribution: Asia: Tigris and Euphrates Basin (Torres, 1991h).

Habitat: Reservoirs, large rivers.

Description: This species has an elongated snake-like body without pelvic fins. Its anal and dorsal fins are elongated and are connected to the caudal fin. The dorsal fin is preceded by numerous spines. Back is dark beige in color, while the head is silver-beige. Body's color is dull brown and belly is a lighter shade of brown. Body may also be marked with brown circular patterns. One to three darker longitudinal zigzag lines that connect to form a distinct reticulated pattern are also seen on the body. Eyes have brown stripes running laterally through them. It grows to a maximum length of 104 cm and 1100 g.

Nutritional Facts

Proximate Composition of Muscle (% WW)

Protein	Fat	Ash	Moisture
19.53	3.33	1.15	75.05

Minerals (mg/kg)

Ca	Na	K	Cu	Mn	Fe	Zn
254.01	650.05	3829.55	0.76	0.17	4.04	11.10

Fatty Acids (% of Total Fatty Acids)

C14:0	2.67
C15:0	0.02
C16:0	19.86
C17:0	0.52
C18:0	3.15
C20:0	0.02
C22:0	0.06
C23:0	0.67
C24:0	6.89
Total SFA	33.87
C14:1n9	0.31
C15:1	0.26
C16:1n9	7.48
C16:1n7	0.88
C17:1n9	0.27
C18:1n9	20.92
C18:1n7	4.92
C20:1n11	0.12
C20:1n9	1.72
C22:1n11	0.03
C22:1n9	1.6
Total MUFA	38.49
C18:2n9	0.47
C18:2n6	3.12
C18:3 n6	0.08
C18:3n3	1.93
C18:4n3	0.66
C20:2n6	0.28
C20:4n6	3.63
C20:3n3	0.57
C20:4n3	0.97
C20:5n3 (EPA)	1.62
C22:6n3 (DHA)	8.41
Total PUFA	21.74
n-3	14.16
n-6	7.11
Unidentified	5.9

Source: Data from Taşbozan et al. (2013a).

Macrognathus aculeatus Bloch 1786

Order: Synbranchiformes

Family: Mastacembelidae

Common name: Lesser spiny eel.

Distribution: Southeast Asia: Malaysia, Thailand, Borneo, and Indonesia.

Habitat: Medium to large-sized rivers; lowland wetlands and peats.

Description: This species has an upper body that is yellow. A black line runs down the middle, and lower body is usually a mix of white and brown. Along the backbone of the eel, the dorsal fin is preceded by numerous isolated small spines that can be raised. The dorsal fin also has many prominent

eyespots along the base. This species can grow up to 35 cm.

Nutritional Facts

Vitamin A and Minerals (per 100 g Raw Cleaned Parts, WW)

Vitamin A (RAE)	Ca[a]	Fe[b]	Zn[b]
90	0.4	2.4	1.2

Total effect of vitamin A is expressed in retinol activity equivalents (RAEs); one retinol equivalent (RE) corresponds to 1 µg retinol, 2 µg β-carotene, 6 µg β-carotene, and 12 µg of α-carotene, γ-carotene, or β-cryptoxanthin in food.

[a] g.

[b] mg.

Source: Data from Thilsted (2012).

ORDER: BELONIFORMES (HALFBEAKS, NEEDLEFISHES)

FAMILY: BELONIDAE (NEEDLEFISHES)

Xenentodon cancila Hamilton 1822

Phylum: Chordata

Class: Actinopterygii

Order: Beloniformes

Family: Belonidae

Common name: Freshwater garfish.

Distribution: Asia: Sri Lanka and India (Reyes, 1993l).

Habitat: Marine, freshwater, brackish; streams and rivers and marshy lowland habitats such as swamps and oxbows; man-made canals and irrigation channels; temporarily inundated floodplains.

Description: Body of this species is very elongated and slightly compressed.

Green-silvery dorsally, grading to whitish below. A silvery band with a dark margin runs along the side. A series of four or five blotches is seen on sides between the pectoral and anal fins in adults. Dorsal and anal fins have dark edges. It reaches a length of 40 cm (Reyes, 1993l).

Nutritional Facts

Proximate Composition (% WW)

Moisture	Ash	Protein	Lipid
79.57	2.02	15.65	2.76

Source: Data from Ahmed et al. (2012).

ORDER: ANGUILLIFORMES (EELS)

FAMILY: ANGUILLIDAE (FRESHWATER EELS)

Anguilla japonica Temminck & Schlegel 1846

Phylum: Chordata

Class: Actinopterygii

Class: Anguilliformes

Family: Anguillidae

Common name: Japanese eel.

Distribution: Asia: Japan to the East China Sea, Taiwan, Korea, China, and northern Philippines (Masuda et al., 1984).

Habitat: Catadromous; freshwater, estuaries and coastal environments, including rivers, streams, and wetlands.

Description: Body of this fleshy and smooth species is elongated, cylindrical anteriorly, compressed posteriorly. Mouth corner is extending to posterior margin of eyes. Lower jaw is slightly longer than upper. Lips are symmetrical. Snout is depressed and stout. Dorsal and anal fins are confluent with caudal fin. This species is plain colored, not marbled or mottled. It grows to a maximum length of 1.5 m and weight of 1.9 kg (FAO, 2014a).

Nutritional Facts

Proximate Composition (% WW)

Moisture	Protein	Lipid	Ash
67.54	17.10	11.45	1.20

Fatty Acids (% WW)

Saturates		38.79
Myristic acid (C14:0)	6.72	
Palmitic acid (C16:0)	24.81	
Stearic acid (C18:0)	7.26	
Monoenes		28.21
Palmitoleic acid (C16:1n7)	7.13	
Vaccenic acid (C18:1n7)	3.45	
Oleic acid (C18:1n9)	15.08	
Eicosenoic acid (C20:1n9)	2.55	
High unsaturates (n-6)		4.13
Linoleic acid (C18:2n6)	1.98	
Docosatetraenoic acid (C22:4n6)	0.23	
Arachidonic acid (C20:4n6)	1.92	
High unsaturates (n-3)		28.48
Linolenic acid (C18:3n3)	0.57	
Eicosapentaenoic acid (C20:5n3)	9.85	
Docosahexaenoicacid (C22:6n3)	18.06	

Source: Data from Seo et al. (2013).

254

Anguilla anguilla Linnaeus 1758 (= *Muraena anguilla*)

Common name: European eel.

Distribution: From North Cape in northern Norway, southward along the coast of Europe, all coasts of the Mediterranean, and on the North African coast (Jacoby and Gollock, 2014).

Habitat: From small streams to large rivers and lakes, and in estuaries, lagoons, and coastal waters.

Description: Body of this species is elongated, cylindrical anteriorly, somewhat compressed posteriorly. Head is rather long. Eye is rounded, and small in young and yellow eels and large in silver eels. Lower jaw is longer than the upper and protruding. Teeth are minute, set in bands in both jaws and in a patch on vomer. Gill openings are small, vertical, and restricted to sides. Dorsal and anal fins are confluent with caudal fin. Dorsal fin originates far behind the pectorals, and anal fin origininates slightly behind anus, well back from origin of dorsal fin. Pectoral fins are small and rounded. Pelvic fins are absent. Lateral line is conspicuous. It has minute, elliptical scales embedded in the skin. There are different stages in the life cycle of European eels. When the larvae (leptocephali) approach the coast, they metamorphose into a transparent larval stage called the "glass eel." These glass eels enter estuaries, and start migrating upstream. After entering fresh water, the glass eels metamorphose into elvers, which are greenish-brown and are miniature versions of the adult eels. As the eel grows, it becomes known as a "yellow eel" due to the brownish-yellow color of its sides and belly. After 5–20 years in fresh water,

the eels become sexually mature. At this stage, their eyes grow larger, their flanks become silver, and their bellies white. At this stage, the eels are known as "silver eels," and they begin their migration back to the sea to spawn (DANAQ, 2014).

Nutritional Facts

Protein Content of the Meat

% of Wet Weight	% of Dry Matter
15.3	55

Essential Amino Acid Composition (g/16 g N of the Protein)

Arginine	9.4
Histidine	2.7
Isoleucine	5.3
Leucine	7.8
Lysine	8.8
Methionine	3.1
Phenylalanine	4.0
Threonine	4.9
Tryptophan	0.9
Valine	5.6

Lipid Content of the Meat

% of Wet Weight	% of Dry Matter
24.5	60

Major Elements in the Muscle (mg/100 g)

Na	K	Mg	Ca	P
65	217	21	17	223

Trace Elements in the Muscle (g/100 g)

Fe	Mn	Se
600	25	31

Source: Data from Steffens et al. (2006).

ORDER: ACIPENSERIFOMES (STURGEON-LIKE FISH)

FAMILY: ACIPENSERIDAE (STURGEON)

Acipenser persicus Borodin 1897

Order: Acipenseriformes

Family: Acipenseridae

Common name: Persian sturgeon.

Distribution: Eurasia: Caspian Basin; eastern Black Sea (Kottelat and Freyhof, 2007).

Habitat: Marine; freshwater; brackish; large and deep rivers.

Description: This species has an elongated, bulky body with a long head. It has the characteristic 8–18 dorsal scutes, 24–50 lateral scutes, and 6–12 ventral scutes. Color of the body is dark blue-black on back with white-cream starry bony platelets to white on ventral side. There are four barbels, which are closer to the end of the snout than the mouth. It grows to a maximum length of 242 cm and weight of 70 kg (Kottelat and Freyhof, 2007; Pond Life).

Amino Acids (g/100 g protein)

Aspartic acid	9.94
Threonine	3.93
Serine	2.74
Glutamic acid	18.4
Proline	4.74
Glycine	4.45
Alanine	5.34
Cysteine	0.81
Valine	5.41
Methionine	2.29
Isoleucine	5.56
Leucine	8.45
Tyrosine	2.89
Phenylalanine	4.30
Histidine	5.19
Lysine	9.42
Arginine	6.08
Triptophan	1.11
Total EAAs	42.97
Total NEAAs	57.85

EAA, essential amino acid; NEAA, nonessential amino acid.

Source: Data from Alipour et al. (2010).

Nutritional Facts

Proximate Composition (g/100 g WW)

Moisture	Protein	Fat	Ash
63.2	21.4	13.1	3.33

Acipenser ruthenus Linnaeus 1758

Order: Acipenseriformes

Family: Acipenseridae

Common name: Sterlet.

Distribution: Eurasia: Black, Azov, and Caspian Seas; Siberia (Birstein, 1993).

Habitat: Rivers migrating between fresh and salt waters.

Description: This species is distinguishable from other European species of sturgeons by the presence of whitish lateral scutes, four long fringed barbels, and an elongated and narrow snout. Ventrals and laterals are very light colored, nearly white or yellowish. It may reach 16 kg in weight and 100–125 cm in length (Keith and Allardi, 2001).

Nutritional Facts

Proximate Composition (g/100 g DM)

Moisture	Protein	Fat	Ash	Cholesterol[a]
75.38	17.54	5.39	0.93	73.59

[a] mg/100 g.

Fatty Acids (g/100 g)

SFA	MUFA	PUFA	PUFA/SFA
32.67	45.97	22.17	0.68

Source: Data from Ljubojevic et al. (2013).

Huso huso **Linnaeus 1758**

Order: Acipenseriformes

Family: Acipenseridae

Common name: Beluga, giant sturgeon, European sturgeon, great sturgeon.

Distribution: Black, Caspian, and Azov Seas as well as the Adriatic (FishBase, 2002).

Habitat: Anadromous (spending at least part of its life in salt water and returning to rivers to breed).

Description: It is the largest sturgeon in the world and the largest European freshwater fish, as it can reach up to 5m in length with a maximum weight of 2072 kg. It has an elongated body shape and a flattened, slightly upturned snout with the mouth located underneath. There are five rows of bony plates (or scutes) that run the length of the body, one along the back, one on each flank, and two on the undersurface. The short, fleshy barbels in front of the mouth are feathered at the ends. Body is predominantly dark gray or greenish, while the belly tends to be white (FishBase, 2002; Cihar, 1991).

Nutritional Facts

Proximate Composition (% WW)

Lipid	Protein	Moisture	Ash
4.25	18.45	70.75	1.90

Fatty Acids (g/100 g of Total Fatty Acids)

SFA	MUFA	PUFA	n-3	n-6
20.67	37.87	33.04	9.04	24.00

Source: Data from Pourshamsian (2012).

6
AMPHIBIANS (FROGS)

FAMILY: DICROGLOSSIDAE

Hoplobatrachus occipitalis Günther 1858

Phylum: Chordata

Class: Amphibia

Order: Anura

Family: Dicroglossidae

Common name: African groove-crowned frog, crowned bullfrog.

Distribution: From southern Mauritania to Ethiopia, south through East Africa to northern Zambia, southern and western Democratic Republic of Congo, Angola, and coastal Congo, Gabon and Cameroon (IUCN SSC Amphibian Specialist Group, 2014).

Habitat: From dry savannas to disturbed forest, using logging roads and rivers to penetrate deep into lowland forest.

Description: A large dorsoventrally flattened ranid frog species with a very warty skin whose eyes and nostrils are dorsally positioned. It has complete webbing between the toes and fingers. Due to numerous glands, the skin is very slippery. It has a broad mouth and the border of the lower jaw has three tooth-like structures. The distinct tympanum is bordered by a bulging supratympanal fold. Males have paired lateral vocal sacs. Females have a size of 127 mm and 235 g. Adult males measure 52–104 mm in length and 24–84 g in weight. The basic color of the body and limbs is a yellowish green, olive, or drab brown. Large dark green to blackish spots, which occasionally form rows, are present on the back (AmphibiaWeb, 2014b).

Nutritional Facts

Proximate Composition (% WW)

Moisture	Protein	Ash	Fats
77.85	19.46	1.28	1.06

Amino Acids (g/100 g Protein or g/16 g N)

Lysine	5.74
Histidine	2.17
Arginine	7.33
Aspartic acid	10.23
Threonine	2.81
Serine	2.82

Glutamic acid	15.37
Proline	2.84
Glycine	4.46
Alanine	3.38
Cystine	1.38
Valine	5.05
Methionine	3.82
Isoleucine	4.12
Leucine	7.29
Tyrosine	3.65
Phenylalanine	5.14

Source: Data from Onadeko et al. (2011).

Hoplobatrachus rugulosus **Wiegmann 1834**

Order: Anura

Family: Dicroglossidae

Common name: Chinese edible frog, East Asian bullfrog, Taiwanese frog, Chinese bullfrog.

Distribution: From central, southern, and southwestern China, including Taiwan, Hong Kong, and Macau to Myanmar through Thailand, Lao People's Democratic Republic, Vietnam, and Cambodia south to the Thai-Malay Peninsula (Diesmos et al., 2014).

Habitat: Paddy fields, irrigation infrastructure, fishponds, ditches, floodplain wetlands, forest pools, and other wet areas.

Description: It is a large, heavy-bodied species with distinct ridges in the dorsal skin. It grows to a size of 12 cm in snout-vent length. Body is olive-brown or gray. Legs are powerful, and feet are fully webbed. Underside is cream, sometimes with dark spotting. Females are larger than males. They are primarily insectivores (Naturalist.org; Reptiles and Amphibians of Bangkok).

Nutritional Facts (WW)

Proximate Composition per 100 g

Protein (g/100 g)	21.82
Fat (g/100 g)	0.67
Amino acids (mg/100 g)	21.07
Taurine (mg/100 g)	692.07
Total FAA (mg/100 g)	2364.78
UFA (% of total FA)	66.59–69.28

FAA, free amino acid; UFA, unsaturated fatty acid.
Source: Data from Ming (2010).

Limnonectes leporinus Andersson 1923

Order: Anura

Family: Dicroglossidae

Common name: Giant river frog.

Distribution: Endemic to Borneo; Brunei, Kalimantan (Indonesia), and Sabah and Sarawak (Malaysia) (Inger et al., 2004.)

Habitat: Along the banks of small to large rocky streams in hilly rain forest.

Description: *Limnonectes* is a genus of fork-tongued frogs collectively known as fanged frogs because they tend to have unusually large teeth, which are small or absent in other frogs. In this species, pectoral girdle is firmisternal.

Sternum is moderately to strongly bifurcate posteriorly. Vomerine teeth are present. The tadpoles develop in quiet side pools of streams.

Nutritional Facts

Proximate Composition (% WW)

Moisture	Ash	Protein	Fat
82.87	0.48	14.69	0.27

Fatty Acid Composition (% of Total Lipids)

SAT	MUFA	PUFA
38.73	22.57	38.24

SAT, saturated fatty acid; MUFA, monounsaturated fatty acid; PUFA, polyunsaturated fatty acid.
Source: Data from Ho et al. (2008).

FAMILY: PTYCHADENIDAE

Ptychadena pumilio Boulenger 1920

Order: Anura

Family: Ptychadenidae

Common name: Medine grassland frog, dwarf grass frog, spotted-throated ridged frog, dwarf rana, little rocket frog.

Distribution: Angola, Benin, Botswana, Cameroon, Central African Republic, Congo, the Democratic Republic of the Congo, Ethiopia, Gabon, Malawi, Mali, Mozambique, Namibia, Nigeria, Senegal, Sierra Leone, South Africa, United Republic of Tanzania, Zambia, Zimbabwe (Amphibia Web, 2014e).

Habitat: Close to water, and especially on the banks of rivers and pools.

Description: A small ranid frog with a moderately pointed snout. Adult males measure 25–32 mm (snout-vent length) and females 25.5–36 mm. Four pairs of symmetrically arranged longitudinal ridges are seen on the back. The third pair, counted from the back, covers just half of the back length, whereas the rest stretch from the eye to the end of the body. All these ridges are continuous. On the flanks, some warts form a short ridge. A short supratympanal fold is present. The tympanum is clearly visible and is slightly concave. Males have paired lateral vocal sacs, enlarged tubercles, and swollen first fingers. Besides showing a dark brown basic color, there are also frogs with clear green basic color and any kind of transitional coloration. A fine line or a broad vertebral band, lined by two ridges, often runs along the vertebra. Both the line and the band may be yellow to red. Most bear a dark orange line (Amphibia Web, 2014e).

Nutritional Facts

Proximate Composition (% WW)

Moisture	Protein	Ash	Fats
78.96	19.79	1.26	0.97

Amino Acids (g/100 g Protein or g/16 g N)	
Lysine	5.91
Histidine	2.33
Arginine	8.19
Aspartic acid	9.65
Threonine	3.03
Serine	3.14
Glutamic acid	14.76
Proline	3.15
Glycine	5.13

Alanine	3.15
Cystine	1.52
Valine	4.82
Methionine	3.51
Isoleucine	5.48
Leucine	8.11
Tyrosine	3.17
Phenylalanine	4.62

Source: Data from Onadeko et al. (2011).

FAMILY: RANIDAE

Rana catesbeiana Shaw 1862 (= *Lithobates catesbeianus*)

Order: Anura

Family: Ranidae

Common name: American bullfrog.

Distribution: Canada, Mexico, United States. Introduced: Argentina, Belgium, Brazil, China, Colombia, Cuba, Dominican Republic, Ecuador, France, Germany, Greece, Haiti, Indonesia, Italy, Jamaica, Japan, Malaysia, Netherlands, Peru, Philippines, Puerto Rico, Singapore, Spain, Taiwan, Thailand, Uruguay, and Venezuela (AmphiaWeb, 2014f).

Habitat: Edge of large, permanent water bodies, such as swamps, ponds, and lakes.

Description: This species is distinguished by lacking dorsolateral folds and having very large tympanums, which are larger than the eye in males. Tips of the fingers and toes are blunt. Webbing is well developed. Skin on the back is rough with random tiny tubercles. There is no dorsolateral fold, but there is a prominent supratympanic fold. Mean snout to vent length for males is 152 mm (range 111–178 mm), and for females it is 162 mm (range 120–183 mm). Males have pigmented nuptial pads. Vocal openings are at the corner of the mouth. Dorsum is green, with or without a net-like pattern of gray or brown on top. Venter is slightly white, sometimes mottled with gray or yellow. Coloration varies widely depending on the locality of the bullfrog (AmphiaWeb, 2014f).

Nutritional Facts

Proximate Composition (% WW)

Moisture	Ash	Protein	Fat
77.73	0.87	18.77	0.64

Fatty Acid Composition (% of Total Lipids)

SAT	MUFA	PUFA
38.91	33.52	27.57

Source: Data from Ho et al. (2008).

Pelophylax esculentus **Linnaeus 1758 (= *Rana esculenta*)**

Order: Anura

Family: Ranidae

Common name: Green frog.

Distribution: Throughout most of western, central, and eastern Europe (Kuzmin et al., 2014).

Habitat: Wetlands such as ponds, channels, ditches, and slow-moving rivers and stream.

Description: In this hybrid species, dorsal coloration is grayish green, olive-green, or green with dark spots that vary in size and number. A light middorsal line from snout to cloaca is usually present. There is no temporal spot. Belly is light, usually with dark spots. Male vocal sacs are positioned behind the mouth angles and are gray. It is a smooth-looking frog (rarely has a warty appearance) and has longer legs than its part ancestor the pool frog (Kent Reptile and Amphibian Group).

Nutritional Facts

Proximate Composition (% WW)

Protein	Moisture	Fat	Ash	Salt
22.2	79.5	1.1	1.8	0.7

Source: Data from Baygar and Ozgur (2010).

Fatty Acids (%)

Mystiric acid	1.13–2.30
Palmitic acid	19.76–23.23
Stearic acid	3.61–6.29
Palmitoleic acid	7.06–13.08
Oleic acid	10.83–16.71
Linoleic acid	6.44–6.71
Gamma-linoleic acid	0.13–0.27
Linoleic acid	2.32–3.37
Cis-11,14,17-Eicosatrienoic acid	4.71–7.92
Cis-5,8,11,14,17-Eicosapentaenoic acid	3.96–6.05
Cis-4,7,10,13,16,19-Docosohexaenoic acid	2.77–6.67

Source: Data from Ozogul et al. (2008).

Minerals (mg/kg)

K	Mg	Fe	Ca
98.24	17.07	3908.1	2337.0

Source: Data from Oduntan et al. (2012).

Pelophylax ridibundus Pallas 1771 (= *Rana ridibunda*)

Order: Anura

Family: Ranidae

Common name: Eurasian marsh frog.

Distribution: Afghanistan, Albania, Armenia, Austria, Azerbaijan, Bahrain, Belarus, Bosnia and Herzegovina, Bulgaria, China, Croatia, Cyprus, Czech Republic, Denmark, Estonia, Finland, France, Georgia, Germany, Greece, Hungary, Islamic Republic of Iran, Iraq, Israel, Italy, Kazakhstan, Kyrgyzstan, Latvia, Lithuania, Luxembourg, the Former Yugoslav Republic of Macedonia, Republic of Moldova, Montenegro, Netherlands, Poland, Romania, Russian Federation, Saudi Arabia, Serbia, Slovakia, Slovenia, Tajikistan, Turkey, Turkmenistan, Ukraine. Introduced: Belgium, Spain, Switzerland, and United Kingdom (AmphibiaWeb, 2014d).

Habitat: Arid areas are largely colonized through river valleys and channels.

Description: Female frog of this species can reach a maximum length of 17 cm, but males remain smaller (around 12 cm). Head is proportionally large and the hind legs are long.

Vomerine teeth are present. Posterior part of the tongue is free and forked. Toes are webbed. Pupil of the eye is horizontal. Snout is moderately sharp. Dorsal coloration has different tints of grayish and green, from entirely gray to green. Large dark dorsal spots vary considerably in size, number, and arrangement. Light middorsal line is often present. There is no temporal spot. Belly is grayish white or grayish yellow with dark spotted or blotched-like pattern, or sometimes without this pattern. Males differ from females by having paired gray vocal sacs behind the mouth angles and nuptial pads on the first finger (AmphibiaWeb, 2014d).

Nutritional Facts

Proximate Composition (% WW)

Protein	Fat	Moisture	Ash
16–20	0.50–0.90	79.00–82.40	0.8–1.00

Vitamins (mg/100 g)

B1	B2	B3	B6
0.042–0.060	0.100–0.170	2.316–3.200	0.045–0.075

Source: Data from Çagiltay et al. (2011).

Hylarana galamensis Duméril and Bibron 1841 (= *Rana galamensis*)

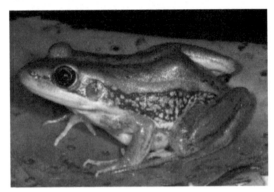

Order: Anura

Family: Ranidae

Common name: Marble-legged frog, golden-backed frog, Lake Galam frog, Galam white-lipped frog, yellow-striped frog.

Distribution: Benin, Burkina Faso, Cameroon, Central African Republic, the Democratic Republic of the Congo, Cote d'Ivoire, Eritrea, Ethiopia, Gambia, Ghana, Guinea-Bissau, Kenya, Malawi, Mali, Mozambique, Nigeria, Senegal, Sierra Leone, Somalia, United Republic of Tanzania, Uganda, Zambia borders (AmphibiaWeb, 2014c).

Habitat: Permanent bodies of water within savannas.

Description: It is large, plump ranid species with comparatively short hind legs. Largest male measures 77.4 mm and largest female reaches 62 mm (snout-vent length). However, males are normally smaller than females. Snout is moderately pointed. Tympanum is clearly visible, reaching about 0.8 of the eye diameter. Males have paired lateral vocal sacs and large glands on the ventral side of the upper arm. Distinct flat, large dorsolateral ridges run from the posterior

eye border to the end of the body. Skin has large flat warts on the flank and posterior back regions, but smooth or finely granulated otherwise. Forelegs are sturdy and toes are without discs. Basic body color varies from drab pale brown to dark brown. Dorsum appears comparatively uniform, showing just several paler yellow spots in the anal region and on the thighs, producing a mottled pattern. Iris is golden to orange. A pale yellow to pale orange stripe runs from the nostril across the eyelid and along the dorsolateral ridge to the body end. On the body, these slightly prominent bands are lined by dark brown to black borders. Upper lip is white, stretching further to the groin as a line of the same color with straight to slightly undulating black borders. Flanks show the same color as the back, bearing several flat yellowish warts with dark borders (AmphibiaWeb, 2014c).

Nutritional Facts

Proximate Composition (g/100 g DM)

Protein	CHO[a]	Fiber	Lipid	Ash
53.74	29.04	1.60	9.52	6.10

[a] Carbohydrate.

Amino Acids (g/100 g Protein)

Lysine	6.93
Histidine	3.13
Arginine	6.55
Asparatic acid	9.66
Threonine	4.33
Serine	5.15
Glutamic acid	13.24
Proline	5.10
Glycine	5.01
Alanine	6.10
Cystine	1.06
Valine	4.82
Methionine	3.00
Isoleucine	4.00
Leucine	7.05

Tyrosine	4.02
Phenylalanine	5.02

Fatty Acids (%)

Lauric acid (12:0)	45.00
Palmitic acid (16:0)	1.23
Stearic acid (18:0)	0.100
Linoleic acid (18:2)	0.00114

Minerals (mg/100 g)

Fe	Mg	Se[a]	Cu	P	Ca
59	429	8065	1.60	14.86	2105

[a] µg.

Source: Data from Muhammad and Ajiboye (2010).

Euphlyctis hexadactylus Lesson 1834 (= *Rana hexadactyla*)

Order: Anura

Family: Ranidae

Common name: Indian green frog, Indian five-fingered frog.

Distribution: Bangladesh, India, Nepal, Sri Lanka (AmphibiaWeb, 2014a).

Habitat: Marshy or damp places; near ponds, agricultural fields, and water of stagnant nature; brackish water.

Description: Size of males is 90 mm and that of females is 130 mm. It has a flattish snout with indistinct canthus rostalis. Tympanum is distinct and equal to or slightly less than the diameter of the eye. First finger is longer than or equal to the second, and toes are fully webbed. Coloration of the body is bright grass green or olive green above with or without a pale yellow vertebral stripe line from snout to vent. Ventrally and on flanks it is white or yellowish white (AmphibiaWeb, 2014a).

Nutritional Facts

Proximate Composition (mg/g DM)

CHO[a]	Glycogen	Protein	Amino Acids	Lipids	Fatty Acids
1.2	3.2	70.8	8.4	38.5	20.4

[a] Carbohydrate.

Source: Data from Janakiram et al. (1983).

Lithobates clamitans Latreille 1801 (= *Rana clamitans*)

Order: Anura

Family: Ranidae

Common name: Bronze frog.

Distribution: Native to the eastern half of the United States and Canada.

Habitat: Shallow freshwater ponds, roadside ditches, lakes, swamps, streams, and brooks.

Description: Adult frogs of this species range from 5 to 10 cm in body length (snout to vent). Typical body weight of this species is from 28 to 85 g. Sexes are sexually dimorphic in a few ways: mature females are typically larger than males; male tympanum is twice the diameter of the eye, whereas in females, tympanum diameter is about the same as that of the eye; and males have bright yellow throats.

Dorsolateral ridges are prominent and seam-like skin folds run down the sides of the back. The dorsum of the frog is generally greenish brown with darker spots or mottling. Center is creamy white—sometimes with dark spots and mottling, especially under the legs. Legs have striped bars.

Nutritional Facts

Proximate Composition (% DM)

DM	Protein	Fat	Ash	Energy[a]
22.5	71.2	10.2	NA	4.80

NA, not available.

[a] kcal/g.

Vitamins (IU/kg) (DM)

Vitamin A	Vitamin E
25,110	82.2

Minerals (DM)

Ca (%)	P (%)	Mg (%)	Na (%)	K (%)	Cu (mg/kg)	Fe (mg/kg)	Zn (mg/kg)	Mn (mg/kg)
4.29	1.87	2.47	0.55	NA	11.2	102.6	100.3	11.5

Source: Data from RodentPro.com (2014).

FAMILY: PIPIDAE

Xenopus muelleri Peters 1844

Order: Anura

Family: Pipidae

Common name: Muller's clawed frog, African clawed frog.

Distribution: Angola, Benin, Botswana, Burkina Faso, Cameroon, Central African Republic, Chad, Congo, the Democratic Republic of the Congo, Gabon, Ghana, Kenya, Malawi, Mozambique, Namibia, Nigeria, South Africa, Sudan, Swaziland, United Republic of Tanzania, Uganda, Zambia, Zimbabwe (AmphibiaWeb, 2014g).

Habitat: Temporary and permanent ponds; streams and rivers.

Description: It is a medium-sized clawed frog with button-like protruding eyes situated dorsally. Body is flattened. Adults measure 38–60 mm (snout-vent length). Subocular tentacles are well developed, reaching about 0.6–1 of the eye diameter. Skin is almost smooth. Starting at the posterior border of the eye, there is a double line of elongate unpigmented flat sensory tubercles (lateral line sense organs). On breeding males, the outer part of the fingers is black. Dermal lobes above the vent are better developed in females. Web of the hind limbs is fully developed. Toes 3–5 have black horny claws. Dorsal parts of body and limbs are drab olive with large black patches that are sometimes vaguely defined. Iris shows a silver-gray glimmer. Venter is whitish to orange yellow. Some darker patches occasionally appear on the throat and on the thighs. Female's belly may be scattered with black spots (AmphibiaWeb, 2014g).

Nutritional Facts

Proximate Composition (% WW)

Moisture	Protein	Ash	Fats
75.60	19.53	1.17	1.81

Amino Acids (g/100 g Protein or g/16 g N)

Lysine	5.08
Histidine	2.27
Arginine	8.02
Aspartic acid	9.80
Threonine	3.11
Serine	2.71
Glutamic acid	14.39
Proline	3.25

Continued

269

Amino Acids (g/100 g Protein or g/16 g N)

Glycine	5.02
Alanine	3.68
Cystine	1.45
Valine	4.73
Methionine	3.46

Isoleucine	4.53
Leucine	7.73
Tyrosine	3.65
Phenylalanine	5.14

Source: Data from Onadeko et al. (2011).

FAMILY: HYLIDAE

Osteopilus septentrionalis A.M.C. Duméril & Bibron 1841

Order: Anura

Family: Hylidae

Common name: Cuban tree frog.

Distribution: Bahamas, Cayman Islands, Cuba. Introduced: Anguilla, Costa Rica, Guadeloupe, Puerto Rico, Turks and Caicos Islands, United States, Virgin Islands, Britain, Virgin Islands (AmphibiaWeb).

Habitat: Tree dwelling and pools.

Description: Frogs of this species are relatively large (140 mm). They have very large toe pads, which are sometimes as large as their tympanum. There is no webbing between the toes on the front legs. However, the rear toes are slightly webbed. Their color is quite variable. They are usually gray to gray green but range to tan brown. These frogs do not have a stripe running through or below their eyes, as some tree frogs do. They have a distinct tarsal fold extending the full length of the tarsus (AmphibiaWeb).

Nutritional Facts

Proximate Composition (% DM)

DM	Protein	Fat	Ash
26.6	NA	4.8	NA

Minerals (DM)

Ca (%)	P (%)
4.79	2.57

Source: Data from RodentPro.com (2014).

REFERENCES

Abbas, F., M. Hafeez-ur-Rehman, M. Ashraf, and K.J. Iqbal. 2013. Body composition of feather back *Notopterus notopterus* and *Rita rita* from Balloki headworks. *Pakistan Journal of Agri-Food and Applied Sciences*, 1: 126–129.

Abdulkarim, B., P.O.J. Bwathondi, and B.L. Benno. 2013. Comparative evaluation of the proximate and mineral compositions of *Stolothrissa tanganicae* and *Limnothrissa miodon* from Lake Tanganyika. *Scientific Journal of Review*, 2: 348–354.

Abimbola, A.O., O.Y. Kolade, A.O. Ibrahim, C.E. Oramadike, and P.A. Ozor. 2010. Proximate and anatomical weight composition of wild brackish *Tilapia guineensis* and *Tilapia melanotheron*. *Internet Journal of Food Safety*, 12: 100–103.

Aboho, S.Y., B.A. Anhwange, and G.A. Ber. 2009. Screening of *Achatina achatina* and *Pila ovata* for trace metals in Makurdi metropolis. *Pakistan Journal of Nutrition*, 8: 1170–1171.

About home. *Puntius conchonius*.

About home. *Trichogaster lalius*.

Adams, S., G.A. Schuster, and C.A. Taylor. 2010a. *Orconectes limosus*. IUCN Red List of Threatened Species. Version 2014.2.

Adams, S., G.A. Schuster, and C.A. Taylor. 2010b. *Orconectes rusticus*. IUCN Red List of Threatened Species. Version 2014.2.

Adebayo-Tayo, B.C., A.A. Onilude, and F.I. Etuk. 2011. Studies on microbiological, proximate mineral and heavy metal composition of freshwater snails from Niger Delta Creek in Nigeria. *Assumption University Journal of Technology*, 14: 290–298.

Adeyeye, E.I., and A.M. Kenni. 2008. The relationship in the amino acid of the whole body, flesh and exoskeleton of common West African freshwater male crab, *Sudanonautes africanus*. *Pakistan Journal of Nutrition*, 7: 748–752.

Adeyeye, E.I., J.O. Olanlokun, and T.O. Falodun. 2010. Proximate and mineral composition of whole body, flesh and exoskeleton of male and female common West African freshwater crab *Sudananautes africanus africanus*. *Polish Journal of Food and Nutrition Sciences*, 60: 213–216.

Ahmad, S.M., U.A. Birnin-Yauri, B.U. Bagudo, and D.M. Sahabi. 2013. Comparative analysis on the nutritional values of crayfish (*Procambarus clarkia*) and some insects. *African Journal of Food Science and Technology*, 4: 9–12.

Ahmed, S., A.F.M.A. Rahman, M.G. Mustafa, M.B. Hossain, and N. Nahar. 2012. Nutrient composition of indigenous and exotic fishes of rainfed waterlogged paddy fields in Lakshmipur, Bangladesh. *World Journal of Zoology*, 7: 135–140.

Akin-Oriola, G., M. Anetekhai, and K. Olowonirejuaro. 2005. Morphometric and meristic studies in two crabs: *Cardiosoma armatum* and *Callinectes pallidus*. *Turkish Journal of Fisheries and Aquatic Sciences*, 5: 85–89.

Al-Bahranay, A.M. 2002. Chemical composition and fatty acid analysis of Saudi Hassawi Rice, *Oryza sativa*, L. *Pakistan Journal of Biological Sciences*, 5: 212–214.

Al-Bassam, K.S. and K.M. Hassan. 2006. Distribution and ecology of recent mollusks in the Euphrates River – Iraq. *Iraqi Bulletin of Geology and Mining*, 2: 57–66.

AlgaeBase. *Anabaena cylindrica*.

AlgaeBase. *Arthrospira maxima*.

AlgaeBase. *Calothrix fusca*.

AlgaeBase. *Chlorella ellipsoidea.*
AlgaeBase. *Chlorella pyrenoidosa.*
AlgaeBase. *Chlorella vulgaris.*
AlgaeBase. *Cladophora* sp.
AlgaeBase. *Cladophora glomerata.*
AlgaeBase. *Diacronema vlkianum.*
AlgaeBase. *Euglena gracilis.*
AlgaeBase. *Gloeocapsa livida.*
AlgaeBase. *Lemanea australis.*
AlgaeBase. *Lemanea fluviatilis.*
AlgaeBase. *Lemanea mamillosa.*
AlgaeBase. *Lemanea torulosa.*
AlgaeBase. *Lyngbya limnetica.*
AlgaeBase. *Nostochopsis lobatus.*
AlgaeBase. *Oscillatoria acuminata.*
AlgaeBase. *Oscillatoria foreaui.*
AlgaeBase. *Paralemanea catenata.*
AlgaeBase. *Phylloderma sacrum.*
AlgaeBase. *Porphyridium cruentum.*
AlgaeBase. *Prasiola japonica.*
AlgaeBase. *Scenedesmus obliquus.*
AlgaeBase. *Synechococcus elongatus.*
Ali, S.S.N., B.K. Tiwari, P. Singh, V. Tripathi, A.A.B. Mahjabin, and A.B. Abidi. 2013. Biochemical variation among some species of pond fishes. *Global Journal of Biology, Agriculture and Health Sciences*, 2: 1–6.
Alinnor, I.J., and C.O. Akalezi. 2010. Proximate and mineral compositions of *Dioscorea rotundata* (white yam) and *Colocasia esculenta* (white cocoyam). *Pakistan Journal of Nutrition*, 9: 998–1001.
Alipour, H.J., B. Shabanpoor, A. Shabani, and A.S. Mahoonak. 2010. Effects of cooking methods on physicochemical and nutritional properties of Persian sturgeon *Acipenser persicus* fillet. *International Aquatic Research*, 2: 15–23.
Allen, D. 2012. *Lepidocephalus guntea.* IUCN Red List of Threatened Species. Version 2014.2.
AmphibiaWeb. 2014. *Euphlyctis hexadactylus.*
AmphibiaWeb. 2014. *Hoplobatrachus occipitalis.*
AmphibiaWeb. 2014. *Hylarana galamensis.*
AmphibiaWeb. 2014. *Pelophylax ridibundus.*
AmphibiaWeb. 2014. *Ptychadena pumilio.*

AmphibiaWeb. 2014. *Rana catesbeiana.*
AmphibiaWeb. 2014. *Xenopus muelleri.*
Anderson, R.A. 1976. Wild rice: Nutritional review. *Cereal Chemistry*, 53: 949–955.
Andini, G. 2009. The potency of freshwater macroalgae *Spirogyra* sp., *Hydrodictyon* sp., *Chara* sp., *Nitella* sp., and *Cladophora* sp. for the source of biodiesel. Final project, degree program in biology, School of Life Sciences and Technology— ITB. Abstract.
Anene, A., O.I. Mba, and O.S. Kalu. 2013. Comparative evaluation of the chemical composition of fillets from two freshwater (*Alestes nurse* and *Oreochromis gallilaeus*) and two brackishwater (*Scomberomorous tritor* and *Psudolithus elongatus*) fish species. *Bioresearch Bulletin*, 1: 73–77.
AnimalBase. *Anisus convexiusculus.*
Animal-world. *Acanthocobitis botia.*
Animal-world. *Rasbora daniconius.*
Animal-world. *Tetraodon lineatus.*
Annandale, N., and S. Kemp. 2013. The crustacea Decapoda of the Lake of Tiberias. *Journal and Proceedings of the Asiatic Society of Bengal* (New Series), 9: 241–258.
Aprodu, I., A. Vasile, G. Gurau, A. Ionscu, and E. Paltenea. 2012. Evaluation of nutritional quality of the common carp (*Cyprinus carpio*) enriched in fatty acids. *Annals of the University Dunarea de Jos of Galati Fascicle VI— Food Technology*, 36: 61–73.
Aquaristik-Elmshorn. *Macrobrachium vollenhovenii.*
Araoye, P.A. 2000. Pectoral spine size in *Synodontis schall* (Teleostei: Mochokidae) from Asa Lake, Ilorin, Nigeria. *Revista de Biología Tropical*, 48: 509–510.
ARKive. *Astacus astacus.*
ARKive. *Cherax destructor.*
ARKive. *Cherax tenuimanus.*
ARKive. *Lissachatina fulica.*
arnobrosi.tripod.com/snails/lanistes.html. *Lanistes libycus.*

Arunachalam, M. 2010. *Neolissochilus hexagonolepis.* IUCN Red List of Threatened Species. Version 2014.2.

Asia Herbs. *Piper sarmentosum.*

Ateş, M., G.Ç. Çakıroğulları, M. Kocabaş, M. Kayım, E. Can, and V. Kızak. 2013. Seasonal variations of proximate and total fatty acid composition of wild brown trout in Munzur River, Tunceli-Turkey. *Turkish Journal of Fisheries and Aquatic Sciences,* 13: 613–619.

Athiyaman, R., and K. Rajendran. 2013. Nutritional value of freshwater prawns, *Macrobrachium scabriculum* (Heller, 1862) and *Macrobrachium idellaidella* (Hilgendorf, 1898). *International Journal of Research in Biological Sciences,* 3: 5–7.

Austin, C.M., C. Jones, and M. Wingfield. 2010. *Cherax quadricarinatus.* IUCN Red List of Threatened Species. Version 2014.2.

Australian Spirulina. TAAU Australia Pty. Ltd. 2014.

Ayeloja, A.A., F.O.A. George, T.O. Dauda, W.A. Jimoh, and M.A. Popoola. 2013. Nutritional comparison of captured *Clarias gariepinus* and *Oreochromis niloticus. International Research Journal of Natural Sciences,* 1: 9–13.

Ayyappan, S. 2014. FAO Fisheries and Aquaculture Department, Rome. http://www.fao.org/fi/common/format/popUpCitation.jsp?type=sourceurl=http%3A//www.fao.org/fishery/country-sector/naso_india/en

Azeroual, A., L. da Costa, P. Lalèyè, and T. Moelants. 2010. *Chrysichthys nigrodigitatus.* IUCN Red List of Threatened Species. Version 2014.2.

Azeroual, A., M. Entsua-Mensah, A. Getahun, and P. Lalèyè. 2010. *Auchenoglanis biscutatus.* IUCN Red List of Threatened Species. Version 2014.

Babu, C.H.S., D.S. Amarnath, and B. Kishor. 2013. Comparison on meat composition of Indian major carps (*Catla catla, Labeo rohita, Cirrhinus mrigala*) and fresh water cat fish (*Pangasius hypophthalmus*) under different treatments. *International Journal of Research in Zoology,* 3: 10–15.

Baby, R.L., I. Hasan, K.A. Kabir, and M.N. Naser. 2010. Nutrient analysis of some commercially important molluscs of Bangladesh. *Journal of Scientific Research,* 2: 390–396.

Bahir, M.M., and D.C.J. Yeo. 2007. The geocarcinucid freshwater crabs of southern India (Crustacea: Decapoda: Brachyura). *Raffles Bulletin of Zoology,* 16: 309–354.

Bailly, N. 1997. FishBase. *Brycon orbignyanus.*

Bailly, N. 1997. FishBase. *Corica soborna.*

Bailly, N. 1997. FishBase. *Gudusia chapra.*

Bailly, N. 1997. FishBase. *Aspius aspius.*

Bailly, N. 1997. FishBase. *Hampala macrolepidota.*

Bailly, N. 1997. FishBase. *Labeo boggut.*

Bailly, N. 1997. FishBase. *Limnothrissa miodon.*

Bailly, N. 1997. FishBase. *Nandus nandus.*

Bailly, N. 1997. FishBase. *Parambassis wolffii.*

Bailly, N. 1997. FishBase. *Puntius gonionotus.*

Bailly, N. 1997. FishBase. *Schizothorax curvifrons.*

Bailly, N. 1997. FishBase. *Stolothrissa tanganicae.*

Bailly, N. 1997. FishBase. *Tenualosa ilisha.*

Bailly, N. 1997. FishBase. *Tetraodon cutcutia.*

Bailly, N. 1997. FishBase. *Tilapia rendalli.*

Baker, A.L. 2012. Phycokey—An image based key to algae (PS Protista) cyanobacteria, and other aquatic objects. University of New Hampshire Center for Freshwater Biology.

Bano, Y. 1977. Seasonal variations in the biochemical composition of *Clarias batrachus*, L. *Proceedings of the Indian Academy of Science,* 85B: 147–155.

Barua, P., and S. Chakraborty. 2011. Proximate composition of egg, stomach content and body composition of Pacu (*Piaractus brachypomus*) collected from aquatic environment of Bangladesh. *Current Biotica*, 5: 330–343.

Bashar, M.A. 2010a. BdFISH feature. *Labeo calbasu*.

Bashar, M.A. 2010b. BdFISH feature. *Salmophasia bacaila*.

Bashar, M.A. 2010c. BdFISH feature. *Tetraodon cutcutia*.

Bashar, M.A. 2011. BdFISH feature. *Puntius ticto*.

Bashar, M.A. 2013. BdFISH feature. *Salmostoma phulo*.

Baygar, T., and N. Ozgur. 2010. Sensory and chemical changes in smoked frog (*Rana esculenta*) leg during cold storage (4°C±1). *Journal of Animal and Veterinary Advances*, 9: 588–593.

Becker, E.W. 1994. *Microalgae: Biotechnology and Microbiology*. Cambridge University Press, Cambridge.

Becker, E.W. 2006. Micro-algae as a source of protein: Research review paper. In *Microalgae: Biotechnology and Microbiology*. Cambridge University Press, Cambridge, 293 pp.

Begum, M., and M.H. Minar. 2012. Comparative study about body composition of different shell fish and Ilisha commonly available in Bangladesh. *Trends in Fisheries Research*, 1: 38–42.

Begum, A., M.H. Minar, Md. Sarower, E. Mahuj, and M. Begum. 2013. Monthly variation of biochemical composition of gonia (*Labeo gonius*) collected from Bangladeshi water. *International Journal of Life Science and Pharma Research*, 2: 227–232.

Beklevk, G., A. Polat, and F. Ozoul. 2005. Nutritional value of sea bass (*Dicentrarchus labrax*) fillets during frozen (–18°C) storage. *Turkish Journal of Veterinary and Animal Sciences*, 29: 891–895.

Bhagowati, A.K., and B.K. Ratha. 1982. Biochemical composition and nutritive values of three species of hillstream fish belonging to the genus *Garra* from north-eastern India. *Proceedings of the National Academy of Sciences, India*, 48: 67–72.

Bhosale, R., J. Rout, and B. Chagule. 2012. The ethnobotanical study of an edible freshwater red alga, *Lemanea fluviatilis* (L.) C.Ag. from Manipur, India. *Ethnobotany Research and Applications*, 10: 69–76.

Big Fishes of the World. *Tor khudree*. http://bigfishesoftheworld.blogspot.in

Bilgin, S., and Z.U.C. Fidanbaş. 2011. Nutritional properties of crab (*Potamon potamios* Olivier, 1804) in the Lake of Eğirdir (Turkey). *Pakistan Veterinary Journal*, 31: 239–243.

Bindu, L. 2006. Life history traits of *Etroplus suratensis* (Bloch). In Captive breeding protocols of two potential cultivable fishes, *Etroplus suratensis* (Blotch) and *Horabagrus brachysoma* (Gunther) endemic to Western Ghat region, Kerala. PhD thesis, Mahatma Gandhi University, Kerala, pp. 45–79.

Binohlan, C.B. 1990a. FishBase. *Barbus grypus*.

Binohlan, C.B. 1990b. FishBase. *Dicentrarchus labrax*.

Binohlan, C.B. 1990c. FishBase. *Silurus triostegus*.

Biodiversity India. *Oryza rufipogon*.

BIOTIC. *Nucella lapillus*. Biological Traits Information Catalogue.

Birstein, V.J. 1993. Sturgeons and paddlefishes: Threatened fishes in need of conservation. *Conservation Biology*, 7: 773–787.

Bishop, C.T., G.A. Adams, and E.O. Hughes. 1954. A polysaccharide from the blue-green alga, *Anabaena cylindrica*. *Canadian Journal of Chemistry*, 32: 999–1004.

Bluegreen Foods AFA. 2001–2014. Bluegreen algae products.

Borba, M.R., D.M. Fracalossi, L.E. Pezzato, D. Menoyo, and J.M. Bautis. 2003. Growth, lipogenesis and body composition of piracanjuba (*Brycon orbignyanus*) fingerlings fed different dietary protein and lipid concentrations. *Aquatic Living Resources*, 16: 362–369.

Boulenger, G.A. 1899. A revision of the genera and species of fishes of the family Mormyridae. In *Proceedings of the General Meetings for Scientific Business of the Zoological Society of London*, pp. 775–821.

Boyd, C.E. 1968a. Evaluation of some common aquatic weeds as possible feedstuffs. *Hyacinth Control Journal/Journal of Aquatic Plant Management*, 7: 26–27.

Boyd, C.E. 1968b. Fresh-water plants: A potential source of protein. *Economic Botany*, 22: 359–363.

Britz, R. 1996. *American Museum of Natural History*, 3181.

Brown, D.S. 2002. *Freshwater snails of Africa and their medical importance*. CRC Press, Boca Raton, FL, 608 pp.

Budha, P.B., and B.A. Daniel. 2010a. *Lamellidens jenkinsianus*. IUCN Red List of Threatened Species. Version 2014.2.

Budha, P.B., and A. Madhyastha. 2010b. *Pila virens*. IUCN Red List of Threatened Species. Version 2014.2.

Budha, P.B., J. Dutta, and B.A. Daniel. 2010. *Bellamya bengalensis*. IUCN Red List of Threatened Species. Version 2014.2.

Budha, P.B., A. Madhyastha, and J. Dutta. 2010. *Pila globosa*. IUCN Red List of Threatened Species. Version 2014.2.

Burchardt, L., S. Balcerkiewicz, M. Kokocinski, S. Samardakiewicz, and S. Adamski. 2006. Occurrence of *Haematococcus pluvialis* flowtow emend. Wille in a small artificial pool on the university campus of the Collegium Biologicum in Poznan (Poland). *Biodiversity: Research and Conservation*, 1–2: 163–166.

Çagiltay, F., N. Erkan, D. Tosun, and A. Selçuk. 2011. Chemical composition of the frog legs (*Rana ridibunda*). *Fleischwirtschaft International* 26: 78–81.

CalorieSlism. *Corbicula fluminea*.

Capuli, E.E. 1991. FishBase. *Trichogaster trichopterus*.

Casal, C.M.V. 1995a. FishBase. *Coregonus albula*.

Casal, C.M.V. 1995b. FishBase. *Esomus longimanus*.

Casal, C.M.V. 1995c. FishBase. *Orecohromis aureus*.

Celik, M., M.A. Gokce, N. Basusta, A. Kucukgulmez, O. Tasbozani, and S.S. Tabakoglu. 2008. Nutritional quality of rainbow trout (*Oncorhynchus mykiss*) caught from the Atatürk Dam Lake in Turkey. *Journal of Muscle Foods*, 19: 50–61.

Center for Aquatic and Invasive Plants, University of Florida. *Ipomoea aquatic.*

Center for Aquatic and Invasive Plants, University of Florida. *Vallisneria americana.*

Center for Aquatic and Invasive Plants, University of Florida. *Zizania aquatica.*

Chaki, N. 2013. BdFISH feature. *Corica soborna.*

Chaudhry, S. 2010. *Chitala chitala*. IUCN Red List of Threatened Species, Version 2014.2.

Chen, D.W., M. Zhang, and S. Shrestha. 2007. Compositional characteristics and nutritional quality of Chinese mitten crab (*Eriocheir sinensis*). *Food Chemistry*, 103: 1343–1349.

Chima, J.U., and E.N.T. Akobundu. 2010. Proximate composition of processed freshwater snail (*Pila ovata*) meat as affected by salting, fermentation, and frying. *Journal of Sustainable Agriculture and the Environment*, 12: 150–156.

Chudiwal, A.K., D.P. Jain, and R.S. Somani. 2010. *Alpingia galanga* Willd.—An overview of phytopharmacological properties. *Indian Journal of Natural Products and Resources*, 1: 143–149.

Chumchal, M. 2002. *Cyprinus carpio*. Animal Diversity Web. (Accessed August 5, 2014). http://animal-diversity.org/accounts/Cyprinus_carpio/

Cihar, J. 1991. *Freshwater fish*. Aventinum Publishing, Prague.

Collaa, L.M., T.E. Bertolin, and J.A.V. Costa. 2004. Fatty acids profile of *Spirulina platensis* grown under different temperatures and nitrogen concentrations. *Zeitschrift für Naturforschung*, 59: 55–59.

Crandall, K.A. 2010a. *Procambarus clarkii*. IUCN Red List of Threatened Species. Version 2014.2.

Crandall, K.A. 2010b. *Procambarus zonangulus*. IUCN Red List of Threatened Species. Version 2014.2.

Crispina, B. 1990. FishBase. *Barbus grypus*.

Crusta-Fauna. http://www.crusta-fauna.org/shrimp-index/macrobrachium-cf-scabriculum

Cumberlidge, N. 1994. Identification of *Sudanonautes aubryi* (H. Milne-Edwards, 1853) (Brachyura: Potamoidea: Potamonautidae) from West and Central Africa. *Zeitschrift für Angewandte Zoologie*, 80: 225–241.

Cumberlidge, N. 2008a. *Potamon potamios*. IUCN Red List of Threatened Species. Version 2014.2.

Cumberlidge, N. 2008b. *Spiralothelphusa hydrodroma*. IUCN Red List of Threatened Species. Version 2014.2.

Cumberlidge, N. 2008b. *Sudanonautes africanus*. IUCN Red List of Threatened Species. Version 2014.2.

Cumberlidge, N. 2008d. *Travancoriana schirnerae*. IUCN Red List of Threatened Species. Version 2014.2.

CyanoDB.cz. *Gloeocapsa livida*.

Dabrowski, T., E. Kolakowski, H. Wawreszuk, and C. Choroszucha. 1966. Studies on chemical composition of American crayfish (*Orconectes limosus*) meat as related to its nutritive value. *Journal of the Fisheries Research Board of Canada*, 23: 1653–1662.

Dadheech, P.K., D.A. Casamatta, P. Casper, and L. Krienitz. 2013. *Phormidium etoshii* sp. nov. (Oscillatoriales, Cyanobacteria) described from the Etosha Pan, Namibia, based on morphological, molecular and ecological features. *Fottea, Olomouc*, 13: 235–244.

D'Agaro, E. 2006. Utilization of water cress (*Nasturtium officinale* L.) in noble crayfish (*Astacus astacus*) feeding. *Bulletin Francais de la Peche et de la Pisciculture*, 380–381: 1255–1260.

Daget, J. 1984. Citharinidae. In *Checklist of the freshwater fishes of Africa (CLOFFA)*, ed. J. Daget, J.P. Gosse, and D.F.E. Thys van den Audenaerde. Vol. 1. ORSTOM, Paris and MRAC, Tervuren, pp. 212–216.

Dahanukar, N. 2010a. *Chanda nama*. IUCN Red List of Threatened Species. Version 2014.2.

Dahanukar, N. 2010b. *Pseudambassis baculis*. IUCN Red List of Threatened Species. Version 2014.2.

Dahanukar, N. 2010c. *Puntius ticto*. IUCN Red List of Threatened Species. Version 2014.2.

Damiani, M.C., C.A. Popovicha, D. Constenla, and P.I. Leonard. 2010. Lipid analysis in *Haematococcus pluvialis* to assess its potential use as a biodiesel feedstock. *Bioresource Technology*, 101: 3801–3807.

DANAQ. 2014. The Danish Aquaculture Development Group. FAO Fisheries and Aquaculture Department, Rome. Online.

Darshan, A., P.C. Mahanta, A. Barat, and P. Kumar. 2013. Redescription of the striped catfish *Mystus tengara* (Hamilton, 1822) (Siluriformes: Bagridae), India. *Journal of Threatened Taxa*, 5: 3536–3541.

Das, M.K., M.K. Bandyopadhyay, A.P. Sharma, S.K. Paul, and S. Bhomick. 2012. *Piscine diversity of River Brahmani—A checklist*. Bulletin 175. Central Inland Fisheries Research Insititute (ICAR), 59 pp.

Dasi, D., and T.C. Nag. 2008. Morphology of adhesive organ of the snow trout *Schizothorax richardsonii* (Gray, 1832). *Italian Journal of Zoology*, 75: 361–370.

De Grave, S. 2013a. *Macrobrachium macrobrachion.* IUCN Red List of Threatened Species. Version 2014.2.

De Grave, S. 2013b. *Macrobrachium malcolmsonii.* IUCN Red List of Threatened Species. Version 2014.2.

De Grave, S., J. Shy, D. Wowor, and T. Page. 2013a. *Macrobrachium rosenbergii.* IUCN Red List of Threatened Species. Version 2014.2.

De Grave, S., D. Wowor, and X. Cai. 2013b. *Macrobrachium scabriculum.* IUCN Red List of Threatened Species. Version 2014.2.

De Grave, S., D. Wowor, and W. Klotz. 2013c. *Macrobrachium rude.* IUCN Red List of Threatened Species. Version 2014.2.

Dejana, T., M. Zoran, P. Radivoj, M. Milan, S. Danka, V. Danijela, and S. Aurelija. 2013. Changes in the proximate and fatty acid composition in carp meat during the semi intensive farming. *Tehnologija mesa,* 54: 39–47.

Deka, S.J., and G.C. Sarma. 2011. Taxonomical studies of Oscillatoriaceae (Cyanophyta) of Goalpara District, Assam, India. *Indian Journal of Fundamental and Applied Life Sciences,* 1: 22–35.

Desai, V.R. 2003. Synopsis of biological data on the tor mahseer *Tor tor* (Hamilton, 1822). FAO, Rome.

Devanji, S., S. Matai, L. Si, S. Barik, and A. Nag. 1993. Chemical composition of two semi-aquatic plants for food use. *Plant Foods for Human Nutrition,* 44: 11–16.

Devi, A.R.S., and M.K. Smija. 2013. Indian analysis of dietary value of the soft tissue of the freshwater crab *Travancoriana schirnerae. Journal of Applied Research,* 3: 45–49.

Devi, C.B., N.K.S. Singh, N.R. Singh, N.R. Singh, M.S.A. Chakraborty, and S.S. Ram. 2011. Trace elements in nungsham, the red edible algae of Manipur. *International Journal of Applied Biology and Pharmaceutical Technology,* 2: 198–203.

Devi, K.R.R., and N. Dahanukar. 2013. *Salmophasia bacaila.* IUCN Red List of Threatened Species. Version 2014.2.

Devi, W.S., and Ch. Sarojnalini. 2012. Impact of different cooking methods on proximate and mineral composition of *Amblypharyngodon mola* of Manipur. *International Journal of Advanced Biological Research,* 2: 641–645.

Dey, R.A. 2007. *Handbook on Indian freshwater molluscs.* Zoological Survey of India.

Diesmos, A., P.P. van Dijk, R. Inger, Dj. Iskandar, M.W.N. Lau, Z. Ermi, L. Shunqing, G. Baorong, L. Kuangyang, Y. Zhigang, G. Huiqing, S. Haitao, and C. Wenhao. 2004. *Hoplobatrachus rugulosus.* IUCN Red List of Threatened Species. Version 2014.2.

Dinakaran, G.K., P. Soundarapandian, and A.K. Tiwary. 2010. Nutritional status of edible palaemonid prawn, *Macrobrachium Scabriculum* (Heller, 1862). *European Journal of Applied Sciences,* 2: 30–36.

djwesten@mst.edu. *Arthrospira platensis* (*Spirulina platensis*).

Dragoş, N., V. Bercea, A. Bica1, B. Drugă, A. Nicoară, and C. Coman. 2010. Astaxanthin production from a new strain of *Haematococcus pluvialis* grown in batch culture. *Annals of the Romanian Society for Cell Biology,* XV: 353–361.

Dunn, J.H., and C.P. Wolk. 1970. Composition of the cellular envelopes of *Anabaena cylindrica. Journal of Bacteriology,* 103: 153–158.

Ecocrop FAO. *Dioscorea rotundata.*

Ecology Asia. *Osteochilus vittatus.*

Edsman, L., L. Füreder, F. Gherardi, and C. Souty-Grosset. 2010. *Astacus astacus.* IUCN Red List of Threatened Species. Version 2014.2.

Effiong, B.N., and J.O. Fakunle. 2011. Proximate and mineral composition of some commercially important fishes in Lake Kainji, Nigeria. *Journal of Basic and Applied Scientific Research*, 1: 2497–2500.

Effiong, B.N., and J.O. Fakunle. 2012. Proximate and mineral content of traditional smoked fish species from Lake Kainji, Nigeria. *Bulletin of Environment, Pharmacology and Life Sciences*, 1: 43–45.

Effiong, B.N., and I. Mohammed. 2008. Effect of seasonal variation in the nutrient composition in selected fish species in Lake Kainji-Nigeria. *Nature and Science*, 6: 1–5.

Egonmwan, R.I. 2008. The ecology and habits of *Tympanotonus fuscatus* var. *radula* (L.) (Prosobranchia: Potamididae). *Journal of Biological Sciences*, 8: 186–190.

Ehigiator, F.A.R., and I.M. Nwangwu. 2011. Comparative studies of the proximate composition of three body parts of two freshwater prawns' species from Ovia River, Edo State, Nigeria (OS). *Australian Journal of Basic and Applied Sciences*, 5: 2899–2903.

Ehigiator, F.A.R., and E.A. Oterai. 2012. Chemical composition and amino acid profile of a Caridean prawn (*Macrobrachium vollenhovenii*) from Ovia River and tropical periwinkle (*Tympanotonus fuscatus*) from Benin River, Edo State, Nigeria. *International Journal of Research and Reviews in Applied Sciences*, 11: 162–167.

Elden Project. *Etlingera elatoir*.

Elegbede, I.O., and H.A. Fashina-Bombata. 2013. Proximate and mineral compositions of common crab species [*Callinectes pallidus* and *Cardisoma armatum*] of Badagry Creek, Nigeria. *Poultry, Fisheries and Wildlife Sciences*, 2: 110.

Emmanuel, B.E., C. Oshionebo, and N.F. Aladetohun. 2011. Comparative analysis of the proximate compositions of *Tarpon atlanticus* and *Clarias gariepinus* from culture systems in south-western Nigeria. *African Journal of Food, Agriculture, Nutrition and Development*, 11: 5344–5359.

Eneji, C.A., A.U. Ogogo, C.A. Emmanuel-Ikpeme, and O.E. Okon. 2008. Nutritional assessment of some Nigerian land and water snail species. *Ethiopian Journal of Environmental Studies and Management*, 1: 56–60.

Entri, D. 2013. Detrmination of *Tor tambroides* (Empurau) growth rate using different feed system. Thesis of BSc with Honours (Resource Biotechnology), Universiti Malaysia, Sarawak, 14 pp.

Environmental Institute of Houston Invasive Species Inventory. *Piaractus brachypomus*.

EOL (Encyclopedia of Life). *Morone saxatilis*.

EOL (Encyclopedia of Life). *Oryza sativa*.

Ersoy, B., and H. Şereflişan. 2010. The proximate composition and fatty acid profiles of edible parts of two freshwater mussels. *Turkish Journal of Fisheries and Aquatic Sciences*, 10: 71–74.

ETYFish Project. *Raiamas guttatus*.

Fahad, M.Y.H. 2011a. BdFISH feature. *Channa marulius*.

Fahad, M.Y.H. 2011b. BdFISH feature. *Garra annandalei*.

Fahad, M.Y.H. 2011c. BdFISH feature. *Garra gatyla*.

Fahad, M.F.H. 2011c. BdFISH feature *Labeo boggut*.

Fahad, M.F.H. 2011e. BdFISH feature. *Labeo pangusia*.

Fahad, M.F.H. 2011f. BdFISH feature *Labeo gonius*.

Fahad, M.F.H. 2012. BdFISH feature. *Barilius bendelisis*.

FAO. *Commelina benghalensis*.

FAO. 2007. *The state of world aquaculture and fisheries 2006*. Food and Agriculture Organization of the United Nations (FAO), Fisheries and Aquaculture Department, Rome, Italy.

FAO. 2014a. Cultured Aquatic Species Information Programme. *Anguilla japonica*.

FAO. 2014b. Species fact sheets. *Abramis brama.*

FAO. 2014c. Species fact sheets. *Channa striata.*

FAO. 2014d. Species fact sheets. *Dicentrarchus labrax.*

FAO. 2014e. Species fact sheets. *Pomatomas saltatrix.*

FAO. 2014f. Species fact sheets. *Pseudolithus elongatus.*

FAO. 2014g. Species fact sheets. *Sander lucioperca.*

FAO/FAOSTAT. 2012. FAO Agriculture Department, Agricultural Production and Livestock Primary.

Fasahat, P., K. Muhammad, A. Abdullah, and W. Ratnam. 2012. Proximate nutritional composition and antioxidant properties of *Oryza rufipogon*, a wild rice collected from Malaysia compared to cultivated rice, MR219. *Australian Journal of Crop Science*, 6: 1502–1507.

Fawole, O.O., M.A. Ogundiran, T.A. Ayandiran, and O.F. Olagunju. 2007. Proximate and mineral composition in some selected fresh water fishes in Nigeria. *Internet Journal of Food Safety*, 9: 52–55.

FishBase. *Cyprinus carpio* Linnaeus, 1758.

FishBase. 2002, September. *Huso huso.*

FishBase. 2003, January. Species account.

FishBase, 2005. *Macrognathus pancalus* (Hamilton, 1822).

FishBase. 2013, November. *Oncorhynchus nerka.*

Flora of Zimbabwe. *Marsilea minuta.* Missouri Botanical Garden.

FLORIDATA. *Hedycium coronarium.*

Froese, R., and D. Pauly. 2003a. FishBase. *Esox lucius.*

Froese, R., and D. Pauly. 2003b. FishBase. *Lates niloticus.*

Froese, R., and D. Pauly. 2003c. FishBase. World Wide Web electronic publication.

Froese, R., and D. Pauly. 2005. FishBase. *Oreochromis niloticus.*

Froese, R., and D. Pauly. 2006a. FishBase. *Puntius sophore.*

Froese, R., and D. Pauly. 2006b. FishBase. *Rasboradaniconius.*

Froese, R., and D. Pauly. 2006c. FishBase. *Schizothorax labiatus.*

Froese, R., and D. Pauly. 2006d. FishBase. *Schizothorax plagiostomus.*

Froese, R., and D. Pauly. 2007a. FishBase. *Bagrusbajad.*

Froese, R., and D. Pauly. 2007b. FishBase. *Helostoma temminckii.*

Froese, R., and D. Pauly. 2011a. FishBase. Species of *Hampala.*

Froese, R., and D. Pauly. 2011b. FishBase. Species of *Mormyrus.*

Froese, R., and D. Pauly. 2012. FishBase. *Anabas testudineus.*

Froese, R., and D. Pauly. 2013. FishBase. *Labeo rohita.*

Froese, R., and D. Pauly. 2014a. FishBase. *Channamicropeltes.*

Froese, R., and D. Pauly. 2014b. FishBase. *Heterotis niloticus.*

Froese, R., and D. Pauly. 2014c. FishBase. *Notopterus notopterus.*

Froese, R., and D. Pauly. 2014d. FishBase. *Synodontis membranacea.*

Galib, S.M. 2008. BdFISH feature. *Clupisoma atherinoides.*

Galib, S.M. 2010a. BdFISH feature. *Amblypharyngodon mola.*

Galib, S.M. 2010b. BdFISH feature. *Cirrhinus cirrhosus.*

Galib, S.M. 2010c. BdFISH feature. *Osteobrama cotio cotio.*

Galib, S.M. 2011. BdFISH feature. *Pseudambassis beculis.*

Galib, S.M. 2013. BdFISH feature. *Lepidocephalus guntea.*

Ganai, R.A. 2012. Studies on the biochemical composition of some selected freshwater fishes of Kashmir Valley. Thesis of Master of Philosophy (M. Phil.) in Zoology, University of Kashmir, Srinagar.

Garcia-Guerro, M., L.S. Racotta, and H. Villarreat. 2000. Variation in lipid, protein and carbohydrate content during the embryonic development of the crayfish, *Cherax quadricarinatus* (Decapoda: Parastacidae). *Journal of Crustacean Biology*, 23: 1–6.

279

Garilao, C.V. 1995a. FishBase. *Pentanemus quinquarius*.

Garilao, C.V. 1995b. FishBase. *Pseudolithus elongatus*.

Garilao, C.V. 1995c. FishBase. *Semiplotus manipurensis*.

Gbadamosi, I.T., and O. Okolosi. 2013. Botanical galactogogues: Nutritional values and therapeutic potentials. *Journal of Applied Biosciences*, 61: 4460–4469.

Geelhand, D. 2013a. FishBase. *Brycinus nurse*.

Geelhand, D. 2013b. FishBase. *Heterobranchus bidorsalis*.

Geelhand, D. 2013c. FishBase. *Leptocypris niloticus*.

Geelhand, D. 2013d. FishBase. *Salmo macrostigma*.

Ghamizi, M., A. Jørgensen, T.K. Kristensen, C. Lange, A.S. Stensgaard, and V.D. Damme. 2010. *Pila ovata*. IUCN Red List of Threatened Species. Version 2014.2.

Gherardi, F., and C. Souty-Grosset. 2010. *Astacus leptodactylus*. IUCN Red List of Threatened Species. Version 2014.2.

Ghomi, M.R., A. Dezhabad, M.S. Dalirie, M. Nikoo, S. Toudar, M. Sohrabnejad, and Z. Babaei. 2012. Nutritional properties of kutum, *Rutilus frisii kutum* (Kamensky), silver carp, *Hypophthalmichthys molitrix* (Val.), and rainbow trout, *Oncorhynchus mykiss* (Walbaum), correlated with body weight. *Archives of Polish Fisheries*, 20: 275–280.

Ghomi, M.R., M. Sohrabnejad, and M. Zarei. 2011. Growth rate, proximate composition and fatty acid profile of juvenile kutum *Rutilus frisii kutum* under light/dark cycles. *Jordan Journal of Biological Sciences*, 4: 37–42.

González, S., G.J. Flicka, S.F. O'Keefe, S.E. Duncana, E. McLean, and S.R. Craig. 2006. Composition of farmed and wild yellow perch (*Perca flavescens*). *Journal of Food Composition and Analysis*, 19: 720–726.

Gopakumar, K. 1975. Fatty acid composition of three species of freshwater fishes. *Fisheries Technology*, 12: 21–24.

Gosse, J.P., and D. Paugy. 2003. Citharinidae. In *The fresh and brackish water fishes of West Africa*, ed. D. Paugy, C. Lévêque, and G.G Teugels. Vol. 1. pp. 313–321.

Gouveia, L., A.P. Batista, I. Sousa, A. Raymundo, and N.M. Bandarra. 2008. Microalgae in novel food products. In *Food chemistry research developments*, ed. K.N. Papadopoulos. Nova Science Publishers, chap. 2, pp. 75–112.

Graphics and Web Programming Team. 2003. *Spirogyra*.

Gunlu, A., and N. Gunlu. 2014. Taste activity value, free amino acid content and proximate composition of Mountain trout (*Salmo trutta macrostigma* Dumeril, 1858) muscles. *Iranian Journal of Fisheries Sciences*, 13: 58–72.

Günther, A.C.L.G. 1864. Description of a new species of *Mormyrus*. In *Proceedings of the General Meetings for Scientific Business of the Zoological Society of London*, p. 22.

Gunther, S.J., R.D. Moccia, and D.P. Bureau. 2005. Growth and whole body composition of lake trout (*Salvelinus namaycush*), brook trout (*Salvelinus fontinalis*) and their hybrid, F1 splake (*Salvelinus namaycus* × *Salvelinus fontinalis*), from first-feeding to 16 weeks post first-feeding. *Aquaculture*, 249: 195–204.

Hameed, I., and G. Dastagir. 2009. Nutritional analyses of *Rumex hastatus* D. Don, *Rumex dentatus* Linn and *Rumex nepalensis* Spreng. *African Journal of Biotechnology*, 8: 4131–4133.

Harlioglue, A.G., S. Aydin, and O. Yilmaz. 2012. Fatty acid, cholesterol and fat-soluble vitamin composition of cray fish *Astacus leptodactylus*. *Food Science and Technology Journal*, 18: 93–100.

280

Hasan, M.R., and R. Chakrabarti, eds. 2009. Emergent aquatic macrophytes. In *Use of algae and aquatic macrophytes as feed in small-scale aquaculture: A review*. FAO Fisheries and Aquaculture Technical Paper 531. FAO, Rome, pp. 89–93.

Hawaiian Plants and Tropical Flowers. *Etlingera elatoir*.

Healthwithfood.org. *Fagopyrum esculentum*.

Healthwithfood.org. *Fagopyrum tataricum*.

Hei, A., and C.H. Sarojnalini. 2012. Proximate composition, macro and micro mineral elements of some smoke-dried hill stream fishes from Manipur, India. *Nature and Science*, 10: 59–65.

Hendry, A.P. 2000. Proximate composition, reproductive development, and a test for trade-offs in captive sockeye salmon. *Transactions of the American Fisheries Society*, 129: 1082–1095.

Hennig, W. 2013. Superfamilia Unionacea. In *Willi Deutsche Zoologische Gesellschaft*. Walter de Gruyter.

Ho, A.L., C.T. Gooi, and H.K. Pang. 2008. Proximate composition and fatty acid profile of anurans meat. *Borneo Science (Journal of Science and Technology)*, 22: 23–29.

Hoek, C. 1995. *Algae: An introduction to phycology*. Cambridge University Press.

http://www.applesnail.net. *Pila ampullacea*

http://www.applesnail.net. *Pomacea bridgesii*

http://www.flowersofindia.net/catalog/slides/Arrowleaf%20Dock.html. Flowers of India. *Rumex hastatus*.

http://www.missouriplants.com/Yellowalt/Jussiaea_repens_page.html. *Jussiaea repens*.

http://www.oswaldasia.org/species/p/polba/polba_en.html. *Polygonum barbatum*.

http://www.petfish.net. *Pomacea bridgesii*.

http://www.rnzih.org.nz/pages/nppa_057.pdf. *Zizania latifolia*.

http://www.seagrant.umn.edu/exotics/rusty.html. *Orconectes rusticus*.

http://www.sealifebase.fisheries.ubc.ca/summary/Callinectes pallidus.html.

http://www.wirbellose.de. *Macrobrachium vollenhovenii*.

Hudu, N., R.S. Dewi, and R. Ahmad. 2010. Proximate, color and amino acid profile of Indonesian traditional smoked catfish. *Journal of Fisheries and Aquatic Science*, 5: 106–112.

Huner, J.V., H. Kononen, P. Henttonen, O.V. Lindqvist, and J. Jussila. 1996. Proximate analysis of freshwater crayfishes and selected tissues with emphasis on cambarids. *Freshwater crayfish*, XI, 227–234.

India Biodiversity Portal. *Garra lissorhynchus*.

Indrayan, A.K., S. Sharma, D. Durgapal, N. Kumar, and M. Kumar. 2005. Determination of nutritive value and analysis of mineral elements for some medicinally valued plants from Uttaranchal. *Current Science*, 89: 1252–1255.

Inger, R., D. Iskandar, I. Das, R. Stuebing, M. Lakim, P. Yambun, and Mumpuni. 2004. *Limnonectes leporinus*. IUCN Red List of Threatened Species. Version 2014.2.

Islam, M.T., S. Ahmed, M.A. Sultana, A.S. Tumpa, and F.A. Flowra. 2013. Nutritional and food quality assessment of dried fishes in Singra Upazila under Natore District of Bangladesh. *DAMA International*, 2: 14–17.

Ismail, S., M.S. Kamarudin, and E. Ramezani-Fard. 2013. Performance of commercial poultry offal meal as fishmeal replacement in the diet of juvenile Malaysian mahseer, *Tor tambroides*. *Asian Journal of Animal and Veterinary Advances*, 8: 284–292.

IUCN SSC Amphibian Specialist Group. 2014. *Hoplobatrachus occipitalis*. IUCN

Red List of Threatened Species. Version 2014.2.

Iwamoto, K. 1984. Morphological observations on *Prasiola japonica* and related species *Chlorophyta prasiolales*. *Japanese Journal of Phycology*, 32: 269–278.

Iyanova, A.S., A. Khotimchenko, B. Toneva, M.S. Dirnitrova-Konaklieva, and K. Stefanov. 2002. Lipid composition and antioxidative effectivity of different *Spirogyra* species. *Comptes Rendus del' Academie Burgare des Sciences*, 55: 47–50.

Jacoby, D., and M. Gollock. 2014. *Anguilla anguilla*. IUCN Red List of Threatened Species. Version 2014.2.

Jadhav, U. 2009. *Aquaculture technology and development*. PHI Learning Pvt. Ltd, 352 pp.

Jain, A., M. Sundriyal, and R.C. Sundriyal. 2011. Dietary use and conservation concern of edible wetland plants at Indo-Burma hotspot: A case study from Northeast India. *Journal of Ethnobiology and Ethnomedicine*, 7: 29–48.

James, D.G. 2006. The impact of aquatic biodiversity on the nutrition of rice farming households in the Mekong Basin: Consumption and composition of aquatic resources. *Journal of Food Composition and Analysis*, 19: 756–757.

Janakiram, B., Y. Venkateswarlu, G.R. Reddy, and K.S. Babu. 1983. A note on organic composition of muscle of some edible freshwater animals. *Indian Journal of Fisheries*, 30: 175–177.

Jankowska, B., Z. Zakęś, and T. Żmijewski. 2006. The impact of diet on the slaughter yield, proximate composition, and fatty acids profile of fillets of tench (*Tinca tinca* (L.)). *Archives of Polish Fisheries*, 14: 195–211.

Jankowska, B., Z. Zakęś, T. Żmijewski, M. Szczepkowski, and A. Kowalska. 2007. Slaughter yield, proximate composition, and flesh colour of cultivated and wild perch (*Perca fluviatilis* L.). *Czech Journal of Animal Science*, 52: 260–267.

Jarmolowicz, S., and Z. Zakês. 2014. Amino acid profile in juvenile pikeperch (*Sander lucioperca* (L.))— Impact of supplementing feed with yeast extract. *Archives of Polish Fisheries*, 22: 135–143.

Jena, J.K. 2014. FAO Fisheries and Aquaculture Department, Rome. http://www.fao.org/fi/common/format/popUpCitation.jsp?type=sourceurl=http%3A//www.fao.org/fishery/culturedspecies/Catla_catla/en

Jenkins, A., and A. Ali. 2013. *Rasbora daniconius*. IUCN Red List of Threatened Species. Version 2014.2.

Jenkins, A., F.F. Kullander, and H.H. Tan. 2009. *Kryptopterus micronema*. IUCN Red List of Threatened Species. Version 2014.2.

Jha, B.R., and A. Rayamajhi. 2010a. *Heteropneustes fossilis*. IUCN Red List of Threatened Species. Version 2014.2.

Jha, B.R., and A. Rayamajhi. 2010b. *Tor putitora*. IUCN Red List of Threatened Species. Version 2014.2.

Jha, G.N., D. Sarma, T.A. Qureshi, and T. Jha. 2014. Effect of beetroot (*Beta vulgaris*) and apple (*Pyrus malus*) peel fortified diets on growth, body composition and total carotenoid content of *Barilius bendelisis*. In *Aquaculture America*, ICAR, New Delhi, meeting abstract.

John, D.J. 2002. *The freshwater algal flora of the British Isles: An identification guide to freshwater and terrestrial algae.* Vol. 1. Cambridge University Press, 714 pp.

Johnson, D.S. 1973. Notes on some species of the genus *Macrobrachium* (Crustacea: Decapoda: Caridea: Palaemonidae). *Journal of the Singapore National Academy of Science*, 3: 274–291.

Johnson, R.P. 1974. *Synopsis of biological data on Sarotherodon galilaeus*. Fisheries Synopsis 90. FAO, Rome.

Johnston, H.W. 1970. The biological and economic importance of algae. Part 3. Edible algae of fresh and brackish waters. *Tuatara*, 18: 19–34.

Jones, C. 2011. FAO Fisheries and Aquaculture Department, Rome. Online.

Jones, M. 2014. FAO Fisheries and Aquaculture Department, Rome. http://www.fao.org/fi/common/format/popUpCitation.jsp?type=sourceurl=http%3A//www.fao.org/fishery/culturedspecies/Salmo_salar/en

Jørgensen, A., T.K. Kristensen, and A.S. Stensgaard. 2010. *Lanistes libycus*. IUCN Red List of Threatened Species. Version 2014.2.

Kareen, O.A. 2005. FishBase. *Morone saxatilis*.

Keith, P., and J. Allardi. 2001. Atlas des poissons d'eau douce de France. *Patrimoines Naturels*, 47: 1–387.

Kent Reptile and Amphibian Group. *Pelophylax esculentus*.

Keremah, R.I., and G. Amakiri. 2013. Proximate composition of nutrients in fresh adult catfishes: *Chrysichthys nigrodigitatus*, *Heterobranchus bidorsalis* and *Clarias gariepinus* in Yenagoa, Nigeria. *Greener Journal of Agricultural Sciences*, 3: 291–294.

Kesner-Reyes, K. 2002a. FishBase. *Oncorhynchus nerka*.

Kesner-Reyes, K. 2002b. FishBase. *Salmo salar*.

Kesner-Reyes, K. 2002c. FishBase. *Scoberomorus tritor*.

Kew Royal Botanic Garden. *Jussiaea suffruticosa*.

Khuantrairong, T., and S. Traichaiyaporn. 2011. The nutritional value of edible freshwater alga *Cladophora* sp. (Chlorophyta) grown under different phosphorus concentrations. *International Journal of Agriculture and Biology*, 13: 297–300.

Kightlinger, W., K. Chen, A. Pourmir, D.W. Crunkleton, G.L. Price, and T.W. Johannes. 2014. Production and characterization of algae extract from *Chlamydomonas reinhardtii*. *Electronic Journal of Biotechnology*, 17: 14–18.

Köhler, F. 2011. *Corbicula leana*. IUCN Red List of Threatened Species. Version 2014.2.

Komárek, J., C.L. Sant´Anna, M. Bohunická, J. Mareš, G.S. Hentschke, J. Rigonato, and M.F. Fiore. 2013. Phenotype diversity and phylogeny of selected *Scytonema* species (Cyanoprokaryota) from SE Brazil. *Fottea, Olomouc*, 13: 173–200.

Kosygin, L., H. Lilabati, and W. Vishwanath. 2001. Proximate composition of commercially important hill stream fishes of Manipur. *Indian Journal of Fisheries*, 48: 111–114.

Kottelat, M., and J. Freyhof. 2007. *Handbook of European freshwater fishes*. Publications Kottelat, Cornol, Switzerland, 646 pp.

Kucera, P., and P. Marvan. 2004. Taxonomy and distribution of Lemanea and Paralemanea (Lemaneaceae, Rhodophyta) in the Czech Republic. *Preslia, Praha*, 76: 163–174, 2004.

Küpeli, T., H. Altundağ and M. İmamoğlu. 2014. Assessment of trace element levels in muscle tissues of fish species collected from a river, stream, lake, and sea in Sakarya, Turkey. *The Scientific World Journal*, Article ID 496107.

Kuzmin, S., T. Beebee, F. Andreone, B. Anthony, B. Schmidt, A. Ogrodowczyk, V. Ishchenko, N. Ananjeva, N. Orlov, B. Tuniyev, M. Ogielska, Cl. Miaud, J. Loman, D. Cogalniceanu, and T. Kovács. 2014. *Pelophylax esculentus*. IUCN Red List of Threatened Species. Version 2014.2.

Laamanen, M.J., L. Forsström, and K. Sivone. 2002. Diversity of *Aphanizomenon flos-aquae* (cyanobacterium) populations along a Baltic Sea salinity gradient. *Applied and Environmental Microbiology*, 68: 5296–5303.

Lahanov, A.P., R.S. Muzalevskaja, N.V. Shelepina, and V. Gorkova. 2004. Biochemical characteristics of some species of genus *Fagopyrum* Mill. In *Proceedings of the 9th International Symposium on Buckwheat*, Prague, pp. 604–611.

Langer, S., P. Manhas, Y. Bakhtiyar, S. Rayees, and G. Singh. 2013. Studies on the seasonal fluctuations in the proximate body composition of *Paratelphusa masoniana* (Henderson) (female), a local freshwater crab of Jammu region. *Advance Journal of Food Science and Technology*, 5: 986–990.

Laskar, B.A., D. Sarma, and D.N. Das. 2013. Biometrics and sexual dimorphism of *Neolissochilus hexagonolepis* (McClelland). *International Journal of Fisheries and Aquatic Studies*, 1: 8–12.

Lim, T.K. 2013. *Edible Medicinal and Non-Medicinal Plants*. Springer, Dordrecht.

Linn, J.G., E.J. Staba, R.D. Goodrich, J.C. Meiske, and D.E. Otterby. 1975. Nutritive value of dried or ensiled aquatic plants. I. Chemical composition. *Journal of Animal Science*, 41: 601–609.

Little, E.C.S. 1979. II. Water, mineral and protein content and productivity of aquatic plants. FAO Fisheries Technical Paper 187. In *Handbook of utilization of aquatic plants*.

Living World of Molluscs. *Anisus convexiusculus*.

Ljubojevic, D., D. Trbovic, J. Lujic, O. Bjelic-Cabrilo, D. Kostic, N. Novakov, and M. Cirkovic. 2013. Fatty acid composition of fishes from inland waters. *Bulgarian Journal of Agricultural Science*, 19: 62–71.

Lopes-Lima, M., and M.B. Seddon. 2014a. *Unio terminalis*. IUCN Red List of Threatened Species. Version 2014.2.

Lopes-Lima, M., and M.B. Seddon. 2014b. *Unio tigridis*. IUCN Red List of Threatened Species. Version 2014.2.

Lorenz, R.T. 1999. A technical review of *Haematococcus* algae. *NatuRose™ Technical Bulletin*, 060.

LRMW (Landcare Research Manaaki Whenua). *Cladophora glomerata*.

Luna, S.M. 1988a. FishBase. *Abramis brama*.

Luna, S.M. 1988b. FishBase. *Ailia coila*.

Luna, S.M. 1988c. FishBase. *Barilius bendelisis*.

Luna, S.M. 1988d. FishBase. *Catla catla*.

Luna, S.M. 1988e. FishBase. *Cirrhinus cirrhosus*.

Luna, S.M. 1988f. FishBase. *Clarias batrachus*.

Luna, S.M. 1988g. FishBase. *Clupisoma garua*.

Luna, S.M. 1988h. FishBase. *Esomus danricus*.

Luna, S.M. 1988i. FishBase. *Eutropiichthys vacha*.

Luna, S.M. 1988j. FishBase. *Glossogobius giuris*.

Luna, S.M. 1988k. FishBase. *Hypophthalmichthys molitrix*.

Luna, S.M. 1988l. FishBase. *Lota lota*.

Luna, S.M. 1988m. FishBase. *Megalops atlanticus*.

Luna, S.M. 1988n. FishBase. *Neolissochilus hexagonolepis*.

Luna, S.M. 1988o. FishBase. *Neolissochilus stracheyi*.

Luna, S.M. 1988p. FishBase. *Osteobrama cotio cotio*.

Luna, S.M. 1988q. FishBase. *Rasbora tornieri*.

Luna, S.M. 1988r. FishBase. *Sander lucioperca*.

Luna, S.M. 1988s. FishBase. *Schizothorax richardsonii*.

Luna, S.M. 1988t. FishBase. *Sperata aor*.

Madhyastha, A., Budha, P.B., and Daniel, B.A. 2010. *Lamellidens marginalis*. IUCN Red List of Threatened Species. Version 2014.2.

Madkour, F.F., A.E. Kamila, and H.S. Nasra. 2012. Production and nutritive value of *Spirulina platensis* in reduced cost media. *Egyptian Journal of Aquatic Research*, 38: 51–57.

Maigret, J., and B. Ly. 1986. Les Poissons de mer de Mauritanie. *Sci. Nat.,* Compiegne, 213 pp.

Malathi, S., and S. Thippeswamy. 2013. The proximate and mineral compositions of freshwater mussel *Parreysia corrugata* (Mullar, 1774) from Tunga River in the Western Ghats, India. *Global Journal of Biology, Agriculture and Health Sciences,* 2: 165–170.

Manhas, P., S. Langer, and R.K. Gupta. 2013. Biochemical composition and caloric content of *Paratelphusa masoniana* (Henderson), a local freshwater crab, from Jammu waters. *International Journal of Recent Scientific Research,* 4: 658–661.

Mani, S. 2013. *Marsilea minuta.* IUCN Red List of Threatened Species. Version 2014.2.

Manirujjaman, M., M.M.H. Khan, M. Uddin, M. Islam, M. Rahman, M. Khatun, S. Biswas, and M.A. Islam. 2014. Comparison of different nutritional parameters and oil properties of two fish species (*Catla catla* and *Cirrhinus cirrhosus*) from wild and farmed sources found in Bangladesh. *Journal of Food and Nutrition Research,* 2: 47–50.

Manivannan, K., G. Thrumaran, K. Devi, P. Anantharaman, and T. Balasubramanian. 2009. Proximate composition of different groups of seaweeds from Vedalai coastal waters (Gulf of Mannar): Southeast coast of India. *Middle-East Journal of Scientific Research,* 4: 72–77.

Manojkumar, T.G. 2006. Fish habitats and species assemblages in the selected rivers of Kerala and investigation on life history traits of *Puntius carnaticus* (Jerdon, 1849). Thesis submitted to Doctor of Philosophy, Cochin University of Science and Technology, pp. 127–131.

Marine Species Dictionary. *Ozeotelphusa senex senex* (= *Oziotelphusa wagrakarowensis*).

Marine Species Dictionary. *Paralemanea catenata.*

Marine Species Identification Portal. *Pseudotolithus typus.*

MarLIN (Marine Life Information Network). *Nucella lapillus.*

Maryland Department of Natural Resources. *Procambarus zonangulus.*

Masuda, H., K. Amaoka, C. Araga, T. Uyeno, and T. Yoshino. 1984. *The fishes of the Japanese Archipelago.* Vol. 1. Tokai University Press, Tokyo, 437 pp.

Mazumder, M.S.A., M.M. Rahman, A.T.A. Ahmed, M. Begum, and M.A. Hossain. 2008. Proximate composition of some small indigenous fish species in Bangladesh. *International Journal of Sustainable Crop Production,* 3: 18–23.

McAlain, W.R., and R.P. Romaire. 2007. FAO Fisheries and Aquaculture Department, Rome. (Accessed August 23, 2014). http://www.fao.org/fi/common/format/popUpCitation.jsp?type=sourceurl=http%3A//www.fao.org/fishery/culturedspecies/Procambarus_clarkii/en

Mealographer. 2006. *Aplodinotus grunniens.*

MeD India. Medical/health website—Networking for health medicinal plants of Bangladesh. *Enhydra fluctuans.*

Memidex. *Zizania aquatica.*

Menon, A.G.K. 1999. Check list—Fresh water fishes of India. *Records of Zoological Survey of India,* Occasional Paper 175.

MetaMicrobe.com. *Chlamydomonas reinhardtii.*

MicrobeWiki. 2011. *Euglena gracilis.*

Ming, X.M. 2010. A comparative study on meat quality of tiger frog (*Hoplobatrachus rugulosus*) from different regions of China. Master's thesis, Zhegiang Normal University, China.

MinnesotaWildflowers.info (K. Chayka). *Nuphar variegata.*

Missouri Botanical Garden. *Caltha palustris.*

Missouri Botanical Garden. *Carex stricta.*

Missouri Botanical Garden. *Nelumbo nucifera.*

Missouri Botanical Garden. *Typha latifolia.*

Mohamed, H.A.E., R. Al-Maqbaly, and H.M. Mansour. 2010. Proximate composition, amino acid and mineral contents of five commercial Nile fishes in Sudan. *African Journal of Food Science*, 4: 650–654.

Mohanty, B.P., P. Paria, D. Das, S.Ganguly, P. Mitra, A. Verma, S. Sahoo, A. Mohanty, M. Aftabudddin, B.K. Behera, T.V. Sankar, and A.P. Sharma. 2012. Nutrient profile of giant river-catfish. *Sperata seenghala* (Sykes). *National Academy Science Letters*, 35: 155–161.

Mohsin, A.B.M. 2012. BdFISH feature. *Cyprinus carpio communis.*

Mona, M.H., N.S. Geasa, K.M. Sharshar, and E.M. Morsy. 2000. Chemical composition of freshwater crayfish (*Procambarus clarkii*) and its nutritive value. *Egyptian Journal of Aquatic Research*, 4: 19–34.

Monago, C.C., and A.A. Uwakwe. 2009. Proximate composition and in-vitro anti sickling property of Nigerian *Cyperus esculentus* (tiger nut sedge). *Trees for Life Journal*, 4: 2.

Monalisa, K., M.Z. Islam, T.A. Khan, A.T.M. Abdullah, and M.M. Hoque. 2013. Comparative study on nutrient contents of native and hybrid koi (*Anabas testudineus*) and pangas (*Pangasius pangasius, Pangasius hypophthalmus*) fish in Bangladesh. *International Food Research Journal*, 20: 791–797.

Mondal, B., M.R. Rahman, M.J. Alam, A.R. Tarafder, M.A.A. Habib, and M.A. Khaleque. 2005. A study on the culture of *Chlorella ellipsoidea* in various concentrations of unripe tomato juice media. *Pakistan Journal of Biological Sciences*, 8: 823–828.

Mongabay.com. Tropical rainforest conservation and environmental science news.

Mongabay.com. 1994–2013. *Gnathonemus tamandua.*

Mongabay.com. 1994–2013. *Nandus nandus.*

Morey, S. 2014. Ichthyology at the Florida Museum of Natural History.

Muhammad, N.O., and B.O. Ajiboye. 2010. Nutrient composition of *Rana galamensis. African Journal of Food Science and Technology*, 1: 27–30.

Mullahy, J.H. 1952. The morphology and cytology of *Lemanea australis* Atk. *Bulletin of the Torrey Botanical Club*, 79: 471–484.

Munro, I.S.R. 2000. *The marine and freshwater fishes of Ceylon.* Biotech Books. Delhi, India, 349 pp.

Musa, A.S.M. 2009. Nutritional quality components of indigenous freshwater fish species, *Puntius stigma* in Bangladesh. *Bangladesh Journal of Scientific and Industrial Research*, 44: 367–370.

MUSSEL Project. *Lamellidens generosus.* http://www.mussel-project.net/.

Natarajan, A.V., and A.G. Jhingran. 1963. On the biology of *Catla catla* from the river Jamuna. *Proceedings of the National Institute of Sciences of India*, 29: 326–355.

Natarajan, M.V., and A. Srinivasan. 1961. Proximate and mineral composition of freshwater fishes. *Indian Journal of Fisheries*, 8: 422–429.

National Tropical Botanical Garden. *Etlingera elatoir.*

Naturalist.org. *Hoplobatrachus rugulosus.*

Naturalist.org. *Potamogeton amplifolius.*

Nayan, R. 2011a. BdFISH feature. *Ailia coila.*

Nayan, R. 2011b. BdFISH feature. *Ailiichthys punctata.*

Nayan, R. 2011c. BdFISH feature. *Clupisoma garua.*

Nayan, R. 2011d. BdFISH feature. *Eutropiichthys vacha.*

Nayan, R. 2011e. BdFISH feature. *Sperata aor.*

Nayan, R. 2011f. BdFISH feature. *Tor putitora.*

New, M.B. 2014. FAO Fisheries and Aquaculture Department, Rome. (Accessed August 23, 2014). http://www.fao.org/fi/common/

format/popUpCitation.jsp?
type=sourceurl=http%3A//www.
fao.org/fishery/culturedspecies/
Macrobrachium_rosenbergii/en

Ng, P.K.L.1994. A note on the freshwater crabs of the genus *Spiralothelphusa Bott*, 1968 (Crustacea: Decapoda: Brachyura: Parathelphusidae), with description of a new species from Sri Lanka. *Journal of South Asian Natural History*, 1: 27-30.

Ng, X.N., F.Y. Chye, and A.M. Ismail. 2012. Nutritional profile and antioxidative properties of selected tropical wild vegetables. *International Food Research Journal*, 19: 1487–1496.

NNSS. 2014. GB non-native species secretariat for *Orconectes limosus*.

NOAA. 2014. *Oncorhynchus tshawytscha*. Fisheries. Office of Protected Resources.

Noverian, H., A.H. Vayghan, and A.R. Valipour. 2011. Effect of different levels of astaxanthin on shell color and growth indices of freshwater crayfish (*Astacus leptodactylus*, Eschscholtz, 1823). *World Journal of Fish and Marine Sciences*, 3: 269–274.

NSW Department of Primary Industries. *Cherax quadricarinatus*.

Nurhasan, M. 2008. Nutritional composition of aquatic species in Laotian rice field ecosystems: Possible impact of reduced biodiversity. Master's thesis in International Fisheries Management, University of Tromsø, 76 pp.

Nutrition Facts. 2014. *Nelumbo nucifera*.

Nwaoguikpe, R.N., and R. Nwazue. 2010. The phytochemical, proximate and amino acid compositions of the extracts of two varieties of tiger nut (*Cyperus esculentus*) and their effects on sickle cell hemoglobin polymerization. *Journal of Medicine and Medical Sciences*, 1: 543–549.

Obande, R.A., S. Omeji, and I. Isiguzo. 2013. Proximate composition and mineral content of the fresh water snail (*Pila ampullacea*) from River Benue, Nigeria. *IOSR Journal of Environmental Science, Toxicology and Food Technology*, 2: 43–46.

Oduntan, O.O., J.A. Soaga, and A. Jenyo-Oni. 2012. Comparison of edible frog (*Rana esculenta*) and other bush meat types: Proximate composition, social status and acceptability. *Journal of Environmental Research and Management*, 3: 124–128.

Ogungbenle, H.N., and B.M. Omowole. 2012. Chemical, functional and amino acid composition of periwinkle (*Tympanotonus fuscatus* var *radula*) meat. *International Journal of Pharmaceutical Sciences Review and Research*, 13: 128–132.

Olele, N.F. 2012. Nutrient composition of *Gnathonemus tamandua*, *Chrysichtys nigrodigitatus* and *Auchenoglanis biscutatus* caught from River Niger. *Nigerian Journal of Agriculture, Food and Environment*, 8: 21–27.

Olgunoglu, M.P., I.A. Olgunoğlu, and G. Mustafa. 2014. Seasonal variation in major minerals (Ca, P, K, Mg) and proximate composition in flesh of Mesopotamian catfish (*Silurus triostegus* Heckel,1843) from Turkey. *Annual Research and Review in Biology*, 4: 2628–2633.

Oluwafunmilola, F.O., and M. Ogunkoya. 2006. Effect of smoke-drying on the proximate composition of *Tilapia zillii*, *Parachanna obscura* and *Clarias gariepinus* obtained from Akure, Ondo-State, Nigeria. *Animal Research International*, 3: 478–480.

Onadeko, A.B., R.I. Egonmwan, and J.K. Saliu. 2011. Edible amphibian species: Local knowledge of their consumption in southwest Nigeria and their nutritional value. *West African Journal of Applied Ecology*, 19: 67–76.

Onyia, L.U., C. Milam, J.M. Manuand, and D.S. Allison. 2010. Proximate and mineral composition in some freshwater fishes in Upper River Benue, Yola, Nigeria. *Continental Journal of Food Science and Technology*, 4: 1–6.

Ortañez, A.K. 2005a. FishBase. *Aplodinotus grunniens*.

Ortañez, A.K. 2005b. FishBase. *Parachela siamensis*.

Ortañez, A.K. 2005c. FishBase. *Tilapia zillii*.

Osibona, A.O. 2011. Comparative study of proximate composition, amino and fatty acids of some economically important fish species in Lagos, Nigeria. *African Journal of Food Science*, 5: 581–588.

Osibona, A.O., K. Kusemiju, and G.R. Akande. 2006. Proximate composition and fatty acids profile of the African catfish *Clarias gariepinus*. *Journal of Life and Physical Science*, 3(1): 1–5.

Ozogul, F., Y. Ozogul, A.I. Olgunoglu, and E.K. Boga. 2008. Comparison of fatty acid, mineral and proximate composition of body and legs of edible frog, *Rana esculenta*. *International Journal Food Sciences and Nutrition*, 59: 558–565.

Pablico, G.L. 1997a. FishBase. *Hemibagrus nemurus*.

Pablico, G.L. 1997b. FishBase. *Oncorhynchus keta*.

Pablico, G.L. 1997c. FishBase. *Oncorhynchus kisutch*.

Pastorino, G., and G. Darrigan. 2011. *Pomacea bridgesii*. IUCN Red List of Threatened Species. Version 2014.2.

Patole, S.S. 2012. Studies on nutritional values of some Freshwater fish of riverine system from Sakri Tahsil of Dhulia District (M.S.). *International Indexed and Referred Research Journal*, 3: 52–53.

Paul, D.K., R. Islam, and M.A. Sattar. 2013. Physico-chemical studies of lipids and nutrient contents of *Channa striatus* and *Channa marulius*. *Turkish Journal of Fisheries and Aquatic Sciences*, 13: 487–493.

Pavasovic, A., A.J. Anderson, P.B. Mather, and N.A. Richardson. 2007. Influence of dietary protein on digestive enzyme activity, growth and tail muscle composition in redclaw crayfish, *Cherax quadricarinatus* (von Martens). *Aquaculture Research*, 38: 644–652.

Petenuci, M.E., F.B. Stevanato, J.E. La.Visentainer, M. Matsushita, E.E. Garcia, N. Souza, and J.V. Visentainer. 2008. Fatty acid concentration, proximate composition, and mineral composition in fishbone flour of Nile Tilapia. *Archivos Latinoamericanos de Nutricion* (Organo Oficial de la Sociedad Latinoamericana de Nutrición), 58: 87–90.

Peteri, A. 2014. FAO Fisheries and Aquaculture Department, Rome. http://www.fao.org/fi/common/format/popUpCitation.jsp?type=sourceurl=http%3A//www.fao.org/fishery/culturedspecies/Cyprinus_carpio/en

Petfish.net. *Pomacea bridgesii*.

Philippine Medicinal Plants. *Enhydra fluctuans*.

Philips, I.D. 2010. Biological Synopsis of the Rusty Crayfish (*Orconectes rusticus*). Canadian Manuscript Report of Fisheries and Aquatic Sciences, 2923.

Piao, S., and L. Li. 2001. The actuality of produce and exploitation of *Fagopyrum* in China. In *Proceedings of the 8th ISB*, pp. 571–576.

PlanetCatfish.com Home of aquarium catfishes.

PlanetCatfish.com. 1996–2014. *Rita rita*.

PlanetCatfish.com. 1996–2014. *Sperata aor*.

Plants for a Future. 1996–2012a. *Fagopyrum tataricum*.

Plants for a Future. 1996–2012b. *Nasturtium officinale*.

Plants for a Future. 1996–2012c. *Sagittaria rigida*.

PMNH (Pakistan Museum of Natural History). Virtual world of animals.

Poh, K.R.B. 2006. Morphological and genetic identification, and population structure of Kelah (*Tor tambroides*). Masters thesis (Abstract), Universiti Putra Malaysia.

Pond Life (Doctor Who). *Acipenser persicus*.

Pouomogne, V. 2014. FAO Fisheries and Aquaculture Department, Rome. http://www.fao.org/fi/common/format/popUpCitation.jsp?type=sourceurl=http%3A//www.fao.org/fishery/culturedspecies/Clarias_gariepinus/en

Pourshamsian, K. 2012. Fatty acid and proximate composition of farmed great sturgeon (*Huso huso*) affected by thawing methods, frying oils and chill storage. *Advanced Studies in Biology*, 4: 67–76.

Powell, C.B. 1983. Fresh and brackish water shrimps of economic importance in the Niger Delta. In *2nd Annual Conference of the Fisheries Society of Nigeria (FISON)*, Calabar, Nigeria, January 25–27, 1982, pp. 254–285.

Prabhakar, A.K. and S. P. Roy. 2008. Taxonomic diversity of shell fishes of the Kosi Region of North Bihar (India). *The Ecoscan*, 2: 149–156.

Practical Fishkeeping. *Labeo pangusia*. http://www.practicalfishkeeping.co.uk.

Priyadarshani, I., and B. Rath. 2012. Commercial and industrial applications of micro algae—A review. *Journal of Algal Biomass Utilization*, 3: 89–100.

Protasowicki, M., T. Wlasow, M. Rajkowska, M. Polna and A. Bernard. 2013. Metal concentrations in selected organs of crayfish – *Orconectes limosus* and *Pacifastacus leniusculus* from Mazurian Lakes. *Journal of Elementology*, 683–694.

Raghavan, R. 2013. *Tor khudree*. IUCN Red List of Threatened Species. Version 2014.2.

Rahman, A.K.A. 1989. *Freshwater fishes of Bangladesh*. Zoological Society of Bangladesh, Department of Zoology, University of Dhaka.

Rahman, A.K.A. 2005. *Freshwater fishes of Bangladesh*. 2nd ed. Zoological Society of Bangladesh, Dhaka, Bangladesh, 394 pp.

Rainboth, W.J. 1996. *FAO species identification field guide for fishery purposes: Fishes of the Cambodian Mekong*. FAO, Rome, 265 pp.

Rajeshwari, K.R., and M. Rajashekhar. 2011. Biochemical composition of seven species of cyanobacteria isolated from different aquatic habitats of western Ghats, southern India. *International Journal of Current Trends in Science and Technology*, 2: 240–251.

Rakocy, J.E. 2014. FAO Fisheries and Aquaculture Department, Rome. http://www.fao.org/fi/common/format/popUpCitation.jsp?type=sourceurl=http%3A//www.fao.org/fishery/culturedspecies/Oreochromis_niloticus/en

Ramesh, S., R. Rajan, and R. Santhanam. 2013. *Freshwater phytopharmaceutical compounds*. CRC Press, Boca Raton, FL, 236 pp.

Ramezani-Fard, E., M.S. Kamarudin, C.R. Saad, and S.A. Harmin. 2011. Changes over time in muscle fatty acid composition of Malaysian mahseer, *Tor tambroides*, fed different dietary lipid percentage. *African Journal of Biotechnology*, 10: 18256–18265.

Ramin, E. 2013. Comparison of East African and Iran natural feeding condition based on the chemical and biochemical properties of lake algae. *African Journal of Environmental Science and Technology*, 7: 857–861.

Rangappa, T., P.R. Kumar, P. Jaganmohan, and M.S. Reddy. 2012. Studies on the proximal composition of freshwater prawns *Macrobrachium rosenbergii* and *Macrobrachium malcomsonii*. *World Journal of Fish and Marine Sciences*, 4: 218–222.

Rao, S. N.V., and A. Dey. 1989. Freshwater molluscs in aquaculture. In *Handbook of Freshwater Molluscs of India*. Zoological Survey of India: 225–232.

Ratan, P., and P. Kothiyal. 2011. *Fagopyrum esculentum* Moench (common buckwheat) edible plant of Himalayas: A review. *Asian Journal of Pharmacy and Life Science*, 1: 426–442.

Rayamajhi, A., and B.R. Jha. 2010a. *Garra annandalei*. IUCN Red List of Threatened Species. Version 2014.2.

Rayamajhi, A., and B.R. Jha. 2010b. *Garra gotyla*. IUCN Red List of Threatened Species. Version 2014.2.

Rayamajhi, A., B.R. Jha, and C.M. Sharma. 2010. *Tor tor*. IUCN Red List of Threatened Species. Version 2014.2.

Reed, W., J. Burchard, A.J. Hopson, J. Jenness, and I. Yaro. 1967. *Fish and fisheries of Northern Nigeria*. Ministry of Agriculture, Northern Nigeria, 226 pp.

Reptiles and Amphibians of Bangkok, *Hoplobatrachus rugulosus*.

Reyes, R.B. 1993a. FishBase. *Acanthocobitis botia*.

Reyes, R.B. 1993b. FishBase. *Channa marulius*.

Reyes, R.B. 1993c. FishBase. *Channa micropeltes*.

Reyes, R.B. 1993d. FishBase. *Garra mullya*.

Reyes, R.B. 1993e. FishBase. *Labeo bata*.

Reyes, R.B. 1993f. FishBase. *Labeo calbasu*.

Reyes, R.B. 1993g. FishBase. *Lepidocephalichthys thermalis*.

Reyes, R.B. 1993h. FishBase. *Mystus tengara*.

Reyes, R.B. 1993i. FishBase. *Ompok bimaculatus*.

Reyes, R.B. 1993j. FishBase. *Parambassis ranga*.

Reyes, R.B. 1993k. FishBase. *Tilapia guineensis*.

Reyes, R.B. 1993l. FishBase. *Xenentodon cancila*.

Reyes, R.B. 2011. FishBase. *Rutilus frisii*.

Rhode Island Department of Environmental Management. 2009, June. *Corbicula fluminea*. Fact sheet. Office of Water Resources.

Rhode Island Department of Environmental Management. 2010, August. *Nelumbo lutea*. Fact sheet. Office of Water Resources.

Robinson, E.H., M.H. Li, and D.F. Oberle. 2001. Nutrient characteristics of pond-raised channel catfish. *Mississippi Agricultural and Forestry Experiment Station*, 22: 1–5.

Rodent Pro.com 2014.

Sabinsa Corporation. Bacopin. 2001.

Salman, J., and A.J. Nasar. 2013. Total lipids and total protein in two molluscan species as environmental biomarker of pollution in Euphrates River, Iraq. *International Journal of Current Microbiology and Applied Sciences*, 2: 207–214.

Sampang, A.G. 1999a. FishBase. *Pangasianodon hypophthalmus*.

Sampang, A.G. 1999b. FishBase. *Pangasius pangasius*.

Sanciangco, M. 2002. FishBase. *Parachanna obscura*.

Santhanam, R., N. Sukumaran, and P. Natarajan. 1987. *A manual of freshwater aquaculture*. Oxford and IBH Publishing, New Delhi, 193 pp.

Sarma, D., M.S. Akhtar, N.N. Pandey, N. Shahi, B.P. Mohanty, and P.C. Mahanta. 2011. *Nutrient profile and health benefits of coldwater fishes*. Bulletin. 17. Directorate of Coldwater Fisheries Research (Indian Council of Agricultural Research).

Sarojnalini, C.H. 2010. Nutritive values of two indigenous cobitid fishes *Botia berdmorei* and *Lepidocephalus guntea* of Manipur. *The Bioscan (An International Quarterly Journal of Life Sciences)*, 2: 391–396.

Saxena, S.C. 1959. Adhesion apparatus of a hill-stream cyprinid fish, *Garra mullya* (Sykes). *Proceedings of the National Academy of Sciences, India*, 25B: 205–214.

SBSAC (Soft-Bodied Stream Algae of California). *Lemanea fluviatilis*.

SBSAC (Soft-Bodied Stream Algae of California). *Scytonema bohneri*.

Schultres, F.W. 2012. AnimalBase. *Potomida littoralis*.

Scotcat.com. 2012. *Auchenoglanis biscutatus* search results for *Barbus*

barbus. Environment-Agency.gov.uk (Accessed February 11, 2010.)

Seo, J., J. Choi, J. Seo, T. Ahn, W. Chong, S. Kim, H. Cho, and J. Ahn. 2013. Comparison of major nutrients in eels *Anguilla japonica* cultured with different formula feeds or at different farms. *Journal of Fisheries and Aquatic Science*, 16: 85–92.

SeriouslyFish. 2014a. *Hemisynodontis membranaceus*.

SeriouslyFish. 2014b. *Heteopneustes fossilis*.

SeriouslyFish. 2014c. *Lepidocephalichthys thermalis*.

SeriouslyFish. 2014d. *Leptobarbus hoevenii*.

SeriouslyFish. 2014e. *Leptobarbus rubripinna*.

SeriouslyFish. 2014f. *Puntius conchonius*.

SeriouslyFish. 2014g. *Rasbora tornieri*.

SeriouslyFish. 2014h. *Syncrossus berdmorei*.

Shad, M.A., H. Nawaz, M. Hussain, and B. Yousuf. 2011. Proximate composition and functional properties of rhizomes of lotus (*Nelumbo nucifera*) from Punjab, Pakistan. *Pakistan Journal of Botany*, 43: 895–904.

Shafakatullah, N., S. Shetty, R.O. Lobo, and M. Krishnamoorthy. 2013. Nutritional analysis of freshwater bivalves, *Lamellidens* spp. from River Tunga, Karnataka, India. *Research Journal of Recent Sciences*, 2: 120–123.

Shalab, B., J. Akash, and C. Jasmine. 2012. *Trapa natans* (water chestnut): An overview. *International Journal of Pharmacy*, 3: 31–33.

Sharp, J.H. 1969. Blue-green algae and carbonates—*Schizothrix calcicola* and algal stromatolites from Bermuda. *Limnology and Oceanography*, 14: 568–578.

Sharpe, S. 2014. About home. *Devario aequipinnatus*.

Shetty, S., N.C. Tharavathy, R.O. Lobo, and N. Shafakatullah. 2013. Seasonal changes in the biochemical composition of freshwater bivalves, *Parreysia* spp. from Tungabhadra

River, Karnataka. *International Journal of Pharma Sciences and Research*, 4: 94–99.

Shrestha, J. 1994. *Fishes, fishing implements and methods of Nepal*. Smt. M.D. Gupta, Lalitpur Colony, Lashkar (Gwalior), India, 155 pp.

Shukla, S.N. 2005. *Biodiversity of fish species in aquatic ecosystem of Rewa District*. Final Technical Report on Biodiversity of Fish Species in Aquatic Ecosystem of Rewa District. Government Model Science College, Rewa (M.P.), 63 pp.

Singapore Infopedia. *Eleocharis dulcis*.

Singh, L. 2010. *Garra lissorhynchus*. IUCN Red List of Threatened Species. Version 2014.2.

Singh, M.R., and A. Gupta. 2011. The Nutrient content in fresh water red algae (Lemaneaceae, Rhodophyta) from rivers of Manipur, northeast India. *Electronic Journal of Environmental, Agricultural and Food Chemistry*, 10: 2262–2271.

Singh, S., A. Gautam, A. Sharma, and A. Batra. 2010. *Centella asiatica* (L.): A plant with immense medicinal potential but threatened. *International Journal of Pharmaceutical Sciences Review and Research*, 4: 9–17.

Sophia, B., L. Esther, and N. Eviatar. 2009. Rare species *Lemanea fluviatilis* (L.) Agardh (Rhodophyta) from Israel. *Algological Studies*, 132: 75–89.

Sri-aroon, P., and K. Richter. 2012. *Pila ampullacea*. IUCN Red List of Threatened Species. Version 2014.2.

Stanek, M., B. Kupcewicz, J. Dabrowski, and B. Janicki. 2010. Estimation of fat content and fatty acids profile in the meat of spiny-cheek crayfish (*Orconectes limosus* Raf.) from the Brda River and the Lake Goplo. *Journal of Central European Agriculture*, 3: 297–304.

Steffens, W. 2006. Freshwater fish— Wholesome foodstuffs. *Bulgarian Journal of Agricultural Science*, 12: 320–328.

Stickney, R.R. 2014. FAO Fisheries and Aquaculture Department, Rome. Online.

Sultana, S. 2010a. BdFISH feature. *Chanda nama*.

Sultana, S. 2010b. BdFISH feature. *Colisa fasciatus*.

Susan, M. 1988a. FishBase. *Heteopneustes fossilis*.

Susan, M. 1988b. FishBase. *Mystus cavasius*.

Tacon, A.G.J., and M. Metian. 2013. Fish matters: Importance of aquatic foods in human nutrition and global food supply. *Reviews in Fisheries Science*, 21: 22–38.

Tacon, A.G.J., M. Metian, and S.S. De Silva. 2010. Climate change, food security and aquaculture: Policy implications for ensuring the continued green growth and sustainable development of a much needed aquatic food sector. In *Advancing the aquaculture agenda, Workshop of OECD*, April 15–16, 2010, Paris, pp. 52–58.

Talwar, P.K., and A.G. Jhingran. 1991. *Inland fishes of India and adjacent countries*. Vols. 1–2. CRC Press, Boca Raton, FL, 685 pp.

Tao, W.C., and C.M. Taylor. 2011. *Hedyotis* Linnaeus, Sp. Pl. 1: 101. *Flora China*, 19: 147–174.

Taşbozan, O., M.A. Gokce, M. Celik, S.S. Tabakoglu, A. Kucukgulmez, and A. Baasusta. 2013a. Nutritional composition of spiny eel (*Mastacembelus mastacembelus*) caught from the Atatürk Dam Lake in Turkey. *Journal of Applied Biological Sciences*, 7: 78–82.

Taşbozan, O., F. Ozan, C. Erbas, E.Ü.A. Altuğ, and A.A. Adakli. 2013b. Determination of proximate and amino acid composition of five different tilapia species from the Cukurova region (Turkey). *Journal of Applied Biological Sciences*, 7: 17–22.

Tee, E., S.S. Mizura, R. Kuladevan, S.I. Young, S.C. Khor, and S.K. Chin. 1989. Nutrient composition of Malaysian freshwater fishes. *Proceedings of Nutrition Society of Malaysia*, 4: 63–73.

Teugels, G.G. 1986. A systematic revision of the African species of the genus *Clarias* (Pisces, Clariidae). *Annales du Musée Royal de l'Afrique Centrale. (Zoology)*, 247: 1–199.

Texas Parks and Wildlife Foundation. 2014. *Aplodinotus grunniens*.

Thiamdao, S., M. Motham, J. Pekkoh, L. Mungmai, and Y. Peerapornpisal. 2012. *Nostochopsis lobatus* (bluegreen) Wood em. Geitler (Nostocales), edible algae in Northern Thailand. *Chiang Mai Journal of Science*, 39: 119–127.

Thilsted, S.H. 2012. The potential of nutrient-rich small fish species in aquaculture to improve human nutrition and health. In *Farming the waters for people and food: Proceedings of the Global Conference on Aquaculture*, Phuket, Thailand, 2010, pp. 57–73.

Thomaz, S.M., F.A. Esteves, K.J. Murphy, A.M. dos Santos, A. Caliman, and R.D. Guariento. 2009. Aquatic macrophytes in the tropics: Ecology of populations and communities, impacts of invasions and use by man. In *Tropical biology and conservation management*. Vol. IV. Encyclopedia of Life Support Systems (EOLSS), pp. 1252–1280.

Torres, A.G. 1991a. FishBase. *Barbus barbus*.

Torres, A.G. 1991b. FishBase. *Barbus filamentosus*.

Torres, A.G. 1991c. FishBase. *Cirrhinus mrigala*.

Torres, A.G. 1991d. FishBase. *Colisa fasciatus*.

Torres, A.G. 1991e. FishBase. *Cyprinus carpio*.

Torres, A.G. 1991f. FishBase. *Labeo pangusia*.

Torres, A.G. 1991g. FishBase. *Macrognathus pancalus*.

Torres, A.G. 1991h. FishBase. *Mastacembelus mastacembelus*.

Torres, A.G. 1991i. FishBase. *Oncorhynchus gorbuscha*.

Torres, A.G. 1991j. FishBase. *Oncorhynchus mykiss.*

Torres, A.G. 1991k. FishBase. *Osphronemus goramy.*

Torres, A.G. 1991l. FishBase. *Perca fluviatilis.*

Torres, A.G. 1991m. FishBase. *Salmostoma phulo.*

Torres, A.G. 1991n. FishBase. *Salvelinus fontinalis.*

Torres, A.G. 1991o. FishBase. *Sarotherodon galilaeus.*

Torres, A.G. 1991p. FishBase. *Sarotherodon melanotheron.*

Torres, A.G. 1991q. FishBase. *Schizopyge niger.*

Torres, A.G. 1991r. FishBase. *Schizothorax esocinus.*

Torres, A.G. 1991s. FishBase. *Silurus glanis.*

Torres, A.G. 1991t. FishBase. *Sperata seenghala.*

Torres, A.G. 1991u. FishBase. *Tinca tinca.*

Torres, A.G. 1991v. FishBase. *Wallago attu.*

Uddin, P.K.M.M., S.K. Sarkar, and N. Absar. 2012. Analysis on nutrient contents of different parts of puffer fish (*Tetraodon cutcutia*) and its toxicity. *Journal of Biosciences*, 20: 109–114.

Umar, K.J., L.G. Hassan, S.M. Dangoggo, and M.J. Ladan. 2007. Nutritional composition of water spinach (*Ipomoea aquatica* Forsk.). *Journal of Applied Sciences*, 7: 803–809.

Unlu, E., D. Deger, and T. Cicek. 2012. Comparision of morphological and anatomical characters in two catfish species, *Silurus triostegus* Heckel, 1843 and *Silurus glanis* L., 1758 (Siluridae, Siluriformes). *North Western Journal of Zoology*, 8: 119–124.

U.S. Department of Agriculture. National Nutrient Database for Standard Reference. Standard Release 18. *Pomatomas saltatrix.*

U.S. Department of Agriculture. National Nutrient Database for Standard Reference. Standard Release 26.

U.S. Fish and Wildlife Service. *Piaractus brachypomus.* Web version (accessed August 21, 2012).

USGS. 2012a. *Channa punctatus.* Southeast Ecological Science Center.

USGS. 2012b. *Parachanna obscura.* Southeast Ecological Science Center.

USGS. 2013. NAS—Nonindigenous aquatic species. *Piaractus brachypomus.*

Valdestamon, R.R. 2013. FishBase. *Scatophagus argus argus.*

Vandeputte, M., and L. Labbé. 2012. FAO Fisheries and Aquaculture Department, Rome. Online.

van Vuuren, S.J., J. Taylor, C.V. Ginkel, and A. Gerber. 2006. *Easy identification of the most common freshwater algae: A guide for the identification of microscopic algae in South African freshwaters.* North-West University.

Varadharajan, D., and P. Soundarapandian. 2014. Proximate composition and mineral contents of freshwater crab *Spiralothelphusa hydrodroma* (Herbst, 1794) from Parangipettai, South East Coast of India. *Journal of Aquaculture Research and Development*, 5: 217. doi: 10.4172/2155–9546.1000217.

Vinagre, J., A. Rodríguez, M.A. Larraína, and S.P. Aubourg. 2011. Chemical composition and quality loss during technological treatment in coho salmon (*Oncorhynchus kisutch*). *Food Research International*, 44: 1–13.

Virginia Tech, Virginia State University, College of Agriculture and Life Sciences, College of Natural Resources and Environment. 2009.

Vishwanath, W. 2010. *Raiamas guttatus.* IUCN Red List of Threatened Species. Version 2014.2.

Vishwanath, W., and J. Laisram. 2004. Two new species of *Puntius* Hamilton-Buchanan (Cypriniformes: Cyprinidae) from Manipur, India, with an account of *Puntius* species from the state. *Journal of the Bombay Natural History Society*, 101: 130–137.

Wang, Z.P., and Y. Zhao. 2005. Morphological reversion of *Spirulina* (*Arthrospira*) *platensis* (Cyanophyta): From linear to helical. *Journal of Phycology*, 41: 622–628.

Watve, A. 2013. *Oryza rufipogon*. IUCN Red List of Threatened Species. Version 2014.2.

Weber, M.C.W., and L.F. deBeaufort. 1962. *The fishes of the Indo-Australian Archipelago*. Brill Archive, 448 pp.

Weimin, M. 2014. FAO Fisheries and Aquaculture Department, Rome. Online.

Welcomme, R.L. 1988. *International introductions of inland aquatic species*. FAO Fisheries Technical Paper 294.

Williams, J.D., R.S. Butler, G.L. Warren, and N.A. Johnson. 2014. *Freshwater mussels of Florida*. University of Alabama Press, 498 pp.

Wimalasena, S., and M.N.S. Jayasuriya. 1996. Nutrient analysis of some freshwater fish. *Journal of the National Science Council of Sri Lanka*, 24: 21–26.

Wong, M.M. 1964. The effect of processing on the chemical composition and texture of chinese water chestnut, *Eleocharis dulcis*. MSc thesis, Oregon State University, 76 pp.

World Food Programme (WFP). 2012. Hunger stats. World Food Programme, Rome. http://www.wfp.org/hunger/stats

Wünchiers, R. 2002. *Scenedesmus obliquus*. rw@BioWasserstoff.de

www.Animals.com. *Parreysia khadakvaslaensis*.

www.eFloras.org. *Flora of China*. Vol. 5. *Rumex nepalensis*.

www.eFloras.org. *Flora of China*, Vol. 13. *Viola pilosa*.

Yang, Y., and S. Yen. 1997. Notes on *Limnophila* (Scrophulariaceae) of Taiwan. *Botanical Bulletin—Academia Sinica Taipei*, 38: 285–295.

Yeh, W., G. Chen, and Y. Chen. 2004. Identification of a blue-green alga *Arthrospira maxima* using internal transcribed spacer gene sequence. *Australian Occupational Therapy Journal*, 42: 25–37.

Yeoh, H., and P.M. Wong. 1993. Food value of lesser utilised tropical plants. *Food Chemistry*, 46: 239–241.

Yeşilayer, N., and S.N. Genç. 2013. Comparison of proximate and fatty acid compositions of wild brown trout and farmed rainbow trout. *South African Journal of Animal Science*, 43: 89–97.

Yirankinyuki, F., and B.A.D.W. Lamayi. 2013. Proximate and some minerals analysis of *Colocasia esculenta* (Taro) leaves. *Journal of Medical and Biological Sciences*, 3: 8–14.

Yousaf, M., A. Salam, and M. Naeem. 2011. Body composition of freshwater *Wallago attu* in relation to body size, condition factor and sex from southern Punjab, Pakistan. *African Journal of Biotechnology*, 10: 4265–4268.

Yusof, Y.A.M., J.M.H. Basari, N.A. Mukti, R. Sabuddin, A.R. Muda, S. Sulaiman, S. Makpol, and W.Z.W. Ngah. 2011. Fatty acids composition of microalgae *Chlorella vulgaris* can be modulated by varying carbon dioxide concentration in outdoor culture. *African Journal of Biotechnology*, 10: 13536–13542.

Zaglol, N.F., and F. Eltadawy. 2009. Study on chemical quality and nutrition value of fresh water cray fish (*Procambarus clarkii*). *Journal of the Arabian Aquaculture Society*, 4: 1–16.

Zakaria, Z.A., A.M.M. Jais, Y.M. Goh, M.R. Sulaiman, and M.N. Somchit. 2007. Amino acid and fatty acid composition of an aqueous extract of *Channa striatus* (Haruan) that exhibits antinociceptive activity. *Clinical and Experimental Pharmacology and Physiology*, 34: 198–204.

Zakia, I.A. 2010. BdFISH feature. *Puntius sophore*.

Zakia, I.A. 2014. BdFISH feature. *Cyprinus carpio communis*.

Zanariah, J., and A.N. Rehan. 1988. The consumption, proximate and amino acid composition of local freshwater fish. *Mardi Research Bulletin*, 16: 109–116.

Zhou, Y., Y.H. Shen, C. Zhang, and W.D. Zhang. 2007. Chemical constituents of *Bacopa monnieri*. *Chemistry of Natural Compounds*, 43: 355–357.

INDEX